Rust Web Development with Rocket

A practical guide to starting your journey in Rust web development using the Rocket framework

Karuna Murti

BIRMINGHAM—MUMBAI

Rust Web Development with Rocket

Associate Group Product Manager: Pavan Ramchandani

Publishing Product Manager: Aaron Tanna

Senior Editor: Mark Dsouza

Content Development Editor: Divya Vijayan

Technical Editor: Shubham Sharma

Copy Editor: Safis Editing

Project Coordinator: Rashika Ba

Proofreader: Safis Editing

Indexer: Pratik Shirodkar

Production Designer: Nilesh Mohite

Marketing Coordinators: Anamika Singh and Marylou De Mello

First published: June 2022

Production reference: 1240622

Published by Packt Publishing Ltd.

Livery Place

35 Livery Street

Birmingham

B3 2PB, UK.

ISBN 978-1-80056-130-4

www.packt.com

To my family, Ing, Ping, and Ling, for their 7 years of patience. To my mom, Tjong Njoek Fa, who always supported me. To my dad, Lestoro, who introduced me to the wonderful world of books, reading, and writing.

– Karuna Murti

Contributors

About the author

Karuna Murti started out as a network engineer in Indonesia back in 2005. After that, he continued his career as a web developer, engineering manager, and chief engineer at a consulting company, using various languages such as PHP, Ruby, JavaScript, Fortran, and Python. He also worked with web applications and frameworks such as Magento, WordPress, and Ruby on Rails. In 2015, after establishing his own consulting company, he moved to Japan and now works as a technical lead at Merpay, one of the leading mobile payment apps in Japan. He is helping the company to build a cashless society by leading a team that connects millions of users with thousands of financial institutions in Japan. Even though he uses Go as the main language in his day job, Rust and Rocket have been a passion of his ever since their release.

I want to thank Tomohiro Kato for checking the outline of this book. Many thanks to Robert Jerovšek for reviewing this book.

About the reviewer

Robert Jerovšek is an experienced software engineer who enjoys challenges and has built various types of applications, ranging from mobile to web, but with a particular focus on developing robust backend systems for fast-growing services. He also has experience in management and is currently working as an engineering manager in Japan. Previously, he worked as a unit lead in South Korea and was the CTO of an EdTech start-up in Spain.

Table of Contents

3

Rocket Requests and Responses

4

Building, Igniting, and Launching Rocket

5

Designing a User-Generated Application

Part 2: An In-Depth Look at Rocket Web Application Development

6

Implementing User CRUD

7

Handling Errors in Rust and Rocket

8

Serving Static Assets and Templates

9
Displaying Users' Post

10
Uploading and Processing Posts

11
Securing and Adding an API and JSON

Part 3: Finishing the Rust Web Application Development

12
Testing Your Application

13

Launching a Rocket Application

14

Building a Full Stack Application

15

Improving the Rocket Application

Index

Other Books You May Enjoy

Preface

Rocket is one of the first web frameworks in the Rust programming language. Rocket provides complete tooling to build web applications, such as tools for routing the requests and the strong typing of the incoming requests, and middleware to manipulate incoming requests and outgoing responses. Rocket also provides support for templating and connections to various databases.

Rocket is a web framework written in the Rust programming language. As one of the newer programming languages, Rust is designed to be performant and safe. It is easy to create a safe, multithreaded, and asynchronous application. Rust also has a strong foundation in its tooling, documentation, and community packages. All those strengths contribute to Rust's rapid growth in popularity.

This book explores building a complete web application using the Rocket web framework and the Rust programming language. You will be introduced to various techniques to build a web application that can handle incoming requests, store the data in an RDBMS, and generate proper responses back to any HTTP clients.

Who this book is for

We wrote this book to help software engineers who want to learn how to use the Rocket web framework to build web applications. Although not mandatory, basic knowledge of the Rust programming language will help you understand the topics covered easily.

What this book covers

Chapter 1, *Introducing the Rust Language*, introduces the Rust language and the tools to build Rust applications.

Chapter 2, *Building Our First Rocket Web Application*, guides you through creating and configuring a Rocket application.

Chapter 3, *Rocket Requests and Responses*, introduces the Rocket routes, requests, and responses.

Chapter 4, Building, Igniting, and Launching Rocket, explains two important components of Rocket: state and fairings. State provides reusable objects, and fairings act as the middleware part of the Rocket application. This chapter also explains how to connect a database to the Rocket application.

Chapter 5, Designing a User-Generated Application, explores the process of designing an application and shows how to use Rust modules to create a more manageable web application.

Chapter 6, Implementing User CRUD, guides you on how to **create, read, update, and delete** (**CRUD**) objects in the Rocket web application and the database behind the Rocket application.

Chapter 7, Handling Errors in Rust and Rocket, explains how to handle errors in Rust and how we can apply error handling in the Rocket application.

Chapter 8, Serving Static Assets and Templates, shows how to serve files (such as CSS files and JS files) using the Rocket web application. You will also learn how to use a template to create a response for the Rocket web application.

Chapter 9, Displaying Users' Post, guides you through Rust generics and how to use generics to display different types of user posts.

Chapter 10, Uploading and Processing Posts, explains asynchronous programming and multithreading in Rust applications and how to apply these to process user uploads in the Rocket web application.

Chapter 11, Securing and Adding an API and JSON, guides you through creating authentication and authorization in a Rocket web application. This chapter also explains how to create JSON API endpoints and how to secure API endpoints with JWT.

Chapter 12, Testing Your Application, introduces you to testing the Rust application and creating an end-to-end test for the Rocket web application.

Chapter 13, Launching a Rocket Application, explains how to configure the production server to serve requests using the Rocket web application. This chapter also explains how to containerize the Rust application using Docker.

Chapter 14, Building a Full Stack Application, explains how to use the Rust programming language to build a frontend WebAssembly application to complement the Rocket web application.

Chapter 15, Improving the Rocket Application, explains how to improve and scale the Rocket web application. This chapter also introduces you to possible alternatives to the Rocket web framework.

To get the most out of this book

You will need a Rust compiler by installing Rustup and the stable toolchain on your computer. You can use Linux, macOS, or Windows. For macOS users, it is recommended to use Homebrew. For Windows users, it is recommended to install Rustup using the Windows Subsystem for Linux (WSL or WSL 2).

All the code in this book has been tested on Arch Linux and macOS, but they should work on other Linux distributions or the Windows operating system as well.

Software/hardware covered in the book	Operating system requirements
Rust programming language, stable toolchain	Windows, macOS, or Linux
Rocket web framework 0.5 pre-released	
PostgreSQL database server	
Yew WebAssembly framework	

Since Rust is a compiled language, various development headers might have to be installed on your computer, for example, `libssl-dev` or `libpq-dev`. Pay attention to error messages when compiling the code sample in this book, and install the required libraries for your operating system and development environment if needed.

Further along in the book, in *Chapter 10*, we are going to process videos using the `FFmpeg` command line.

If you are using the digital version of this book, we advise you to type the code yourself or access the code from the book's GitHub repository (a link is available in the next section). Doing so will help you avoid any potential errors related to the copying and pasting of code.

Download the example code files

You can download the example code files for this book from GitHub at `https://github.com/PacktPublishing/Rust-Web-Development-with-Rocket`. If there's an update to the code, it will be updated in the GitHub repository.

We also have other code bundles from our rich catalog of books and videos available at `https://github.com/PacktPublishing/`. Check them out!

Download the color images

We also provide a PDF file that has color images of the screenshots and diagrams used in this book. You can download it here: `https://packt.link/PUFPv`.

Conventions used

There are a number of text conventions used throughout this book.

`Code in text`: Indicates code words in text, database table names, folder names, filenames, file extensions, pathnames, dummy URLs, user input, and Twitter handles. Here is an example: "Built-in data types, such as `Option` or `Result`, handle null-like behavior in a safe manner."

A block of code is set as follows:

```
fn main() {
    println!("Hello World!");
}
```

When we wish to draw your attention to a particular part of a code block, the relevant lines or items are set in bold:

```
impl super::Encryptable for Rot13 {
    fn encrypt(&self) -> String {
    }
}
```

Any command-line input or output is written as follows:

```
rustup default stable
rustup component add clippy
```

Bold: Indicates a new term, an important word, or words that you see onscreen. For instance, words in menus or dialog boxes appear in **bold**. Here is an example: "Select **System info** from the **Administration** panel."

> Tips or Important Notes
> Appear like this.

Get in touch

Feedback from our readers is always welcome.

General feedback: If you have questions about any aspect of this book, email us at customercare@packtpub.com and mention the book title in the subject of your message.

Errata: Although we have taken every care to ensure the accuracy of our content, mistakes do happen. If you have found a mistake in this book, we would be grateful if you would report this to us. Please visit www.packtpub.com/support/errata and fill in the form.

Piracy: If you come across any illegal copies of our works in any form on the internet, we would be grateful if you would provide us with the location address or website name. Please contact us at copyright@packt.com with a link to the material.

If you are interested in becoming an author: If there is a topic that you have expertise in and you are interested in either writing or contributing to a book, please visit authors.packtpub.com.

Share Your Thoughts

Once you've read *Rust Web Development with Rocket*, we'd love to hear your thoughts! Scan the QR code below to go straight to the Amazon review page for this book and share your feedback.

https://packt.link/r/180056130X

Your review is important to us and the tech community and will help us make sure we're delivering excellent quality content.

Part 1:
An Introduction to the Rust Programming Language and the Rocket Web Framework

In this part, you will learn about the Rust programming language, including the basics of it, how to install it on your operating system, and how to use the Rust tools and package registry. You will also create a Rocket web application and configure, compile, and run it. We will look at installing and including other packages for web applications. We will use `sqlx` for web application connection to relational databases.

This part comprises the following chapters:

- *Chapter 1, Introducing the Rust Language*
- *Chapter 2, Building Our First Rocket Web Application*
- *Chapter 3, Rocket Requests and Responses*
- *Chapter 4, Building, Igniting, and Launching Rocket*
- *Chapter 5, Designing a User-Generated Application*

1
Introducing the Rust Language

Almost every programmer has heard about the **Rust** programming language or even tried or used it. Saying "the Rust programming language" every time is a little bit cumbersome, so let's just call it Rust, or the Rust language from this point forward.

In this chapter, we will talk a little bit about Rust to help you if you are new to this language or as a refresher if you have tried it already. This chapter might also help seasoned Rust language programmers a bit. Later in the chapter, we will learn how to install the Rust toolchain and create a simple program to introduce the features of the Rust language. We will then use third-party libraries to enhance one of our programs, and finally, we will see how we can get help for the Rust language and its libraries.

In this chapter, we're going to cover the following main topics:

- An overview of the Rust language
- Installing the Rust compiler and toolchain
- Writing Hello World
- Exploring Rust crates and Cargo
- Exploring other tools and where to get help

Technical requirements

To follow the content of this book, you will need a computer running a Unix-like operating system such as Linux, macOS, or Windows with Windows Subsystem for Linux (WSLv1 or WSLv2) installed. Don't worry about the Rust compiler and toolchain; we will install it in this chapter if you don't have it installed already.

The code for this chapter can be found at `https://github.com/PacktPublishing/Rust-Web-Development-with-Rocket/tree/main/Chapter01`.

An overview of the Rust language

To build web applications using the **Rocket** framework, we must first learn a bit about the Rust language since Rocket is built using that language. According to `https://www.rust-lang.org`, the Rust language is *"a language empowering everyone to build reliable and efficient software."* It began as a personal project for a programmer named Graydon Hoare, an employee at Mozilla, around 2006. The Mozilla Foundation saw the potential of the language for their product; they started sponsoring the project in 2009 before announcing it to the public in 2010.

Since its inception, the focus of Rust has always been on performance and safety. Building a web browser is not an easy job; an unsafe language can have very fast performance, but programmers working with system languages without adequate safety measures in place can make a lot of mistakes, such as missing pointer references. Rust was designed as a system language and learned many mistakes from older languages. In older languages, you can easily shoot yourself in the foot with a null pointer, and nothing in the language prevents you from compiling such mistakes. In contrast, in the Rust language, you cannot write a code that resulted in null pointer because it will be detected during compile time, and you must fix the implementation to make it compile.

A lot of the Rust language design is borrowed from the functional programming paradigm, as well as from the object-oriented programming paradigm. For example, it has elements of a functional language such as closures and iterators. You can easily make a pure function and use the function as a parameter in another function; there are syntaxes to easily make closures and data types such as `Option` or `Result`.

On the other hand, there are no class definitions, but you can easily define a data type, for example, a **struct**. After defining that data type, you can create a block to implement its methods.

Even though there is no inheritance, you can easily group objects by using **traits**. For example, you can create a behavior and name it the `MakeSound` trait. Then, you can determine what methods should be in that trait by writing the method signatures. If you define a data type, for example, a struct named `Cow`, you can tell the compiler that it implements a `MakeSound` trait. Because you say the `Cow` struct implements the `MakeSound` trait, you have to implement the methods defined in the trait for the `Cow` struct. Sounds like an object-oriented language, right?

The Rust language went through several iterations before a stable version was released (Rust 1.0) on May 15, 2015. Some of the early language design was scrapped before releasing the stable release. At one point, Rust had a class feature but this was scrapped before the stable release because Rust design was changed to have data and behavior separation. You write data (for example, in the form of a `struct` or `enum` type), and then you write a behavior (for example, `impl`) separately. To categorize those `impl` in the same group, we can make a **trait**. So, all the functionality you would want from an object-oriented language can be had thanks to that design. Also, Rust used to have garbage collection, but it was then scrapped because another design pattern was used. When objects get out of scope, such as exiting a function, they are deallocated automatically. This type of automatic memory management made garbage collection unnecessary.

After the first stable release, people added more functionalities to make Rust more ergonomic and usable. One of the biggest changes was **async/await**, which was released in version 1.39. This feature is very useful for developing applications that handle I/O, and web application programming handles a lot of I/O. Web applications have to handle database and network connections, reading from files, and so on. People agree that async/await was one of the most needed features to make the language suitable for web programming, because in async/await, the program doesn't need to make a new thread, but it's also not blocking like a conventional function.

Another important feature is `const fn`, a function that will be evaluated at compile-time instead of runtime.

In recent years, many large companies have started to build a talent pool of Rust developers, which highlights its significance in business.

Why use the Rust language?

So, why should we use the Rust language for web application development? Aren't existing established languages good enough for web development? Here are a few reasons why people would want to use the Rust language for creating web applications:

- Safety
- No garbage collection

- Speed
- Multithreading and asynchronous programming
- Statically typed

Safety

Although writing applications using a system programming language is advantageous because it's powerful (a programmer can access the fundamental building block of a program such as allocating computer memory to store important data and then deallocating that memory as soon as it is not in use), it's very easy to make mistakes.

There's nothing in a traditional system language to prevent a program from storing data in memory, creating a pointer to that data, deallocating the data stored in memory, and trying to access the data again through that pointer. The data is already gone but the pointer is still pointing to that part of the memory.

Seasoned programmers might easily spot such mistakes in a simple program. Some companies force their programmers to use a static analysis tool to check the code for such mistakes. But, as programming techniques become more sophisticated, the complexity of the application grows, and these kinds of bugs can still be found in many applications. High-profile bugs and hacks found in recent years, such as *Heartbleed*, can be prevented if we use a memory-safe language.

Rust is a memory-safe language because it has certain rules regarding how a programmer can write their code. For example, when the code is compiled, it checks the lifetime of a variable, and the compiler will show an error if another variable still tries to access the already out-of-scope data. Ralf Jung, a postdoctoral researcher, already made the first formal verification in 2020 that the Rust language is indeed a safe language. Built-in data types, such as `Option` or `Result`, handle null-like behavior in a safe manner.

No garbage collection

Many programmers create and use different techniques for memory management due to safety problems. One of these techniques is garbage collection. The idea is simple: memory management is done automatically during runtime so that a programmer doesn't have to think about memory management. A programmer just needs to create a variable, and when the variable is not used anymore, the runtime will automatically remove it from memory.

Garbage collection is an interesting and important part of computing. There are many techniques such as reference counting and tracing. Java, for example, even has several third-party garbage collectors besides the official garbage collector.

The problem with this language design choice is that garbage collection usually takes significant computing resources. For example, a part of the memory is still not usable for a while because the garbage collector has not recycled that memory yet. Or, even worse, the garbage collector is not able to remove used memory from the heap, so it will accumulate, and most of the computer memory will become unusable, or what we usually call a **memory leak**. In the **stop-the-world** garbage collection mechanism, the whole program execution is paused to allow the garbage collector to recycle the memory, after which the program execution is resumed. As such, some people find it hard to develop real-time applications with this kind of language.

Rust takes a different approach called **resource acquisition is initialization** (**RAII**), which means an object is deallocated automatically as soon as it's out of scope. For example, if you write a function, an object created in the function will be deallocated as soon as the function exits. But obviously, this makes Rust very different compared to programming languages that deallocate memory manually or programming languages with garbage collection.

Speed

If you are used to doing web development with an interpreted language or a language with garbage collection, you might say that we don't need to worry about computing performance as web development is I/O bound; in other words, the bottleneck is when the application accesses the database, disk, or another network, as they are slower than a CPU or memory.

The adage might be primarily true but it all depends on application usage. If your application processes a lot of JSON, the processing is CPU-bound, which means it is limited by the speed of your CPU and not the speed of disk access or the speed of network connection. If you care about the security of your application, you might need to work with hashing and encryption, which are CPU-bound. If you are writing a backend application for an online streaming service, you want the application to work as optimally as possible. If you are writing an application serving millions of users, you want the application to be very optimized and return the response as fast as possible.

The Rust language is a compiled language, so the compiler will convert the program into machine code, which a computer processor can execute. A compiled language usually runs faster than an interpreted language because, in an interpreted language, there is an overhead when the runtime binary interprets the program into native machine code. In modern interpreters, the speed gap is reduced by using modern techniques such as a **just-in-time** (**JIT**) compiler to speed up the program execution, but in dynamic languages such as Ruby, it's still slower than using a compiled language.

Multithreading and asynchronous programming

In traditional programming, synchronous programming means the application has to wait until CPU has processed a task. In a web application, the server waits until an HTTP request is processed and responded to; only then does it go on to handle another HTTP request. This is not a problem if the application just directly creates responses such as simple text. It becomes a problem when the web application has to take some time to do the processing; it has to wait for the database server to respond, it has to wait until the file is fully written on the server, and it has to wait until the API call to the third-party API service is done successfully.

One way to overcome the problem of waiting is multithreading. A single process can create multiple threads that share some resources. The Rust language has been designed to make it easy to create safe multithreaded applications. It's designed with multiple containers such as `Arc` to make it easy to pass data between threads.

The problem with multithreading is that spawning a thread means allocating significant CPU, memory, and OS resources, or what is colloquially known as being *expensive*. The solution is to use a different technique called **asynchronous programming**, where a single thread is reused by different tasks without waiting for the first task to finish. People can easily write an async program in Rust because it's been incorporated into the language since November 7, 2019.

Statically-typed

In programming languages, a dynamically-typed language is one where a variable type is checked at runtime, while a statically-typed language checks the data type at compile time.

Dynamic typing means it's easier to write code, but it's also easier to make mistakes. Usually, a programmer has to write more unit tests in dynamically-typed languages to compensate for not checking the type at compile time. A dynamically-typed language is also considered more expensive because every time a function is called, the routine has to check the passed parameters. As a result, it's difficult to optimize a dynamically-typed language.

Rust, on the other hand, is statically-typed, so it's very hard to make mistakes such as passing a string as a number. The compiler can optimize the resulting machine code and reduce programming bugs significantly before the application is released.

Now that we have provided an overview of the Rust language and its strengths compared to other languages, let's learn how to install the Rust compiler toolchain, which will be used to compile Rust programs. We'll be using this toolchain throughout this book.

Installing the Rust compiler toolchain

Let's start by installing the Rust compiler toolchain. Rust has three official channels: *stable*, *beta*, and *nightly*. The Rust language uses Git as its version control system. People add new features and bug fixes to the master branch. Every night, the source code from the master branch is compiled and released to the nightly channel. After six weeks, the code will be branched off to the beta branch, compiled, and released to the beta channel. People will then run various tests in the beta release, most often in their CI (Continuous Integration) installation. If a bug is found, the fix will be committed to the master branch and then backported to the beta branch. Six weeks after the first beta branch-off, the stable release will be created from the beta branch.

We will use the compiler from the stable channel throughout the book, but if you feel adventurous, you can use the other channels as well. There's no guarantee the program we're going to create will compile if you use another channel though because people add new features and there might be regression introduced in the new version.

There are several ways to install the Rust toolchain in your system, such as bootstrapping and compiling it from scratch or using your OS package manager. But, the recommended way to install the Rust toolchain in your system is by using `rustup`.

The definition on its website (`https://rustup.rs`) is very simple: "*rustup is an installer for the systems programming language Rust.*" Now, let's try following these instructions to install `rustup`.

Installing rustup on the Linux OS or macOS

These instructions apply if you are using a Debian 10 Linux distribution, but if you are already using another Linux distribution, we're going to assume you are already proficient with the Linux OS and can adapt these instructions suitable to your Linux distribution:

1. Open your terminal of choice.
2. Make sure you have cURL installed by typing this command:

   ```
   curl
   ```

3. If cURL is not installed, let's install it:

   ```
   apt install curl
   ```

   If you are using macOS, you will most likely already have cURL installed.

4. After that, follow the instructions on `https://rustup.rs`:

    ```
    curl --proto '=https' --tlsv1.2 -sSf https://sh.rustup.rs
    | sh
    ```

5. It will then show a greeting and information, which you can customize; for now, we're just going to use the default setup:

    ```
    ...
    1) Proceed with installation (default)
    2) Customize installation
    3) Cancel installation

    >
    ```

6. Type 1 to use the default installation.

7. After that, reload your terminal or type this in the current terminal:

    ```
    source $HOME/.cargo/env
    ```

8. You can confirm whether the installation was successful or not by typing `rustup` in the Terminal and you should see the usage instruction for rustup.

9. Now, let's install the stable Rust toolchain. Type the following in the terminal:

    ```
    rustup toolchain install stable
    ```

10. After the toolchain has been installed into your OS, let's confirm whether we can run the Rust compiler. Type `rustc` in the terminal and you should see the instructions on how to use it.

Installing a different toolchain and components

Right now, we have the stable toolchain installed, but there are two other default channels that we can install: *nightly* and *beta*.

Sometimes, you might want to use a different toolchain for various reasons. Maybe you want to try a new feature, or maybe you want to test regression in your application against an upcoming version of Rust. You can simply install it by using `rustup`:

```
rustup toolchain install nightly
```

Each toolchain has components, some of which are required by the toolchain, such as `rustc`, which is the Rust compiler. Other components are not installed by default, for example, `clippy`, which provides more checks not provided by the `rustc` compiler and gives code style suggestions as well. To install it is also very easy; you can use `rustup component add <component>` as shown in this example:

```
rustup default stable
rustup component add clippy
```

Updating the toolchain, rustup, and components

The Rust toolchain has a regular release schedule of around every three months (six weeks plus six weeks), but sometimes there's an emergency release for a major bug fix or a fix for a security problem. As a result, you sometimes need to update your toolchain. Updating is very easy. This command will also update the components installed in the toolchain:

```
rustup update
```

Besides the toolchain, `rustup` itself might also be updated. You can update it by typing the following:

```
rustup self update
```

Now that we have the Rust compiler toolchain installed in our system, let's write our first Rust program!

Writing Hello World!

In this section, we are going to write a very basic program, *Hello World!*. After we successfully compile that, we are going to write a more complex program to see the basic capabilities of the Rust language. Let's do it by following these instructions:

1. Let's create a new folder, for example, `01HelloWorld`.
2. Create a new file inside the folder and give it the name `main.rs`.

3. Let's write our first code in Rust:

    ```
    fn main() {
        println!("Hello World!");
    }
    ```

4. After that, save your file, and in the same folder, open your terminal, and compile the code using the `rustc` command:

    ```
    rustc main.rs
    ```

5. You can see there's a file inside the folder called `main`; run that file from your terminal:

    ```
    ./main
    ```

6. Congratulations! You just wrote your first `Hello World` program in the Rust language.

Next, we're going to step up our Rust language game; we will showcase basic Rust applications with control flow, modules, and other functionalities.

Writing a more complex program

Of course, after making the `Hello World` program, we should try to write a more complex program to see what we can do with the language. We want to make a program that captures what the user inputted, encrypts it with the selected algorithm, and returns the output to the terminal:

1. Let's make a new folder, for example, `02ComplexProgram`. After that, create the `main.rs` file again and add the `main` function again:

    ```
    fn main() {}
    ```

2. Then, use the `std::io` module and write the part of the program to tell the user to input the string they want to encrypt:

    ```
    use std::io;

    fn main() {
        println!("Input the string you want to encrypt:");

        let mut user_input = String::new();
    ```

```
io::stdin()
    .read_line(&mut user_input)
    .expect("Cannot read input");
println!("Your encrypted string: {}", user_input);
}
```

Let's explore what we have written line by line:

I.   The first line, `use std::io;`, is telling our program that we are going to use the `std::io` module in our program. `std` should be included by default on a program unless we specifically say not to use it.

II. The `let...` line is a variable declaration. When we define a variable in Rust, the variable is immutable by default, so we must add the `mut` keyword to make it mutable. `user_input` is the variable name, and the right hand of this statement is initializing a new empty `String` instance. Notice how we initialize the variable directly. Rust allows the separation of declaration and initialization, but that form is not idiomatic, as a programmer might try to use an uninitialized variable and Rust disallows the use of uninitialized variables. As a result, the code will not compile.

III. The next piece of code, that is, the `stdin()` function, initializes the `std::io::Stdin` struct. It reads the input from the terminal and puts it in the `user_input` variable. Notice that the signature for `read_line()` accepts `&mut String`. We have to explicitly tell the compiler we are passing a mutable reference because of the Rust borrow checker, which we will discuss later in *Chapter 9, Displaying User's Post*. The `read_line()` output is `std::result::Result`, an enum with two variants, `Ok(T)` and `Err(E)`. One of the `Result` methods is `expect()`, which returns a generic type `T`, or if it's an `Err` variant, then it will cause panic with a generic error `E` combined with the passed message.

IV. Two Rust enums (`std::result::Result` and `std::option::Option`) are very ubiquitous and important in the Rust language, so by default, we can use them in the program without specifying `use`.

Next, we want to be able to encrypt the input, but right now, we don't know what encryption we want to use. The first thing we want to do is make a **trait**, a particular code in the Rust language that tells the compiler what functionality a type can have:

1. There are two ways to create a module: create `module_name.rs` or create a folder with `module_name` and add a `mod.rs` file inside that folder. Let's create a folder named `encryptor` and create a new file named `mod.rs`. Since we want to add a type and implementation later, let's use the second way. Let's write this in `mod.rs`:

   ```
   pub trait Encryptable {
       fn encrypt(&self) -> String;
   }
   ```

2. By default, a type or trait is private, but we want to use it in `main.rs` and implement the encryptor on a different file, so we should denote the trait as public by adding the `pub` keyword.

3. That trait has one function, `encrypt()`, which has self-reference as a parameter and returns `String`.

4. Now, we should define this new module in `main.rs`. Put this line before the `fn` main block:

   ```
   pub mod encryptor;
   ```

5. Then, let's make a simple type that implements the `Encryptable` trait. Remember the Caesar cipher, where the cipher substitutes a letter with another letter? Let's implement the simplest one called ROT13, where it converts `'a'` to `'n'` and `'n'` to `'a'`, `'b'` to `'o'` and `'o'` to `'b'`, and so on. Write the following in the `mod.rs` file:

   ```
   pub mod rot13;
   ```

6. Let's make another file named `rot13.rs` inside the `encryptor` folder.

7. We want to define a simple struct that only has one piece of data, a string, and tell the compiler that the struct is implementing the `Encryptable` trait. Put this code inside the `rot13.rs` file:

   ```
   pub struct Rot13(pub String);

   impl super::Encryptable for Rot13 {}
   ```

You might notice we put `pub` in everything from the module declaration, to the trait declaration, struct declaration, and field declaration.

8. Next, let's try compiling our program:

```
> rustc main.rs
error[E0046]: not all trait items implemented, missing:
`encrypt`
 --> encryptor/rot13.rs:3:1
  |
3 | impl super::Encryptable for Rot13 {}
  | ^^^^^^^^^^^^^^^^^^^^^^^^^^^^^^^^^^^^^^ missing
  `encrypt` in implementation
  |
 ::: encryptor/mod.rs:6:5
  |
6 |     fn encrypt(&self) -> String;
  |     ------------------------------------------------
 ------ `encrypt` from trait


error: aborting due to previous error

For more information about this error, try `rustc
--explain E0046`.
```

What is going on here? Clearly, the compiler found an error in our code. One of Rust's strengths is helpful compiler messages. You can see the line where the error occurs, the reason why our code is wrong, and sometimes, it even suggests the fix for our code. We know that we have to implement the super::Encryptable trait for the Rot13 type.

If you want to see more information, run the command shown in the preceding error, rustc --explain E0046, and the compiler will show more information about that particular error.

9. We now can continue implementing our Rot13 encryption. First, let's put the signature from the trait into our implementation:

```
impl super::Encryptable for Rot13 {
    fn encrypt(&self) -> String {
    }
}
```

The strategy for this encryption is to iterate each character in the string and add 13 to the char value if it has a character before 'n' or 'N', and remove 13 if it has 'n' or 'N' or characters after it. The Rust language handles Unicode strings by default, so the program should have a restriction to operate only on the Latin alphabet.

10. On our first iteration, we want to allocate a new string, get the original String length, start from the zeroeth index, apply a transformation, push to a new string, and repeat until the end:

```
fn encrypt(&self) -> String {
    let mut new_string = String::new();
    let len = self.0.len();
    for i in 0..len {
        if (self.0[i] >= 'a' && self.0[i] < 'n') ||
        (self.0[i] >= 'A' && self.0[i] < 'N') {
            new_string.push((self.0[i] as u8 + 13) as
            char);
        } else if (self.0[i] >= 'n' && self.0[i] <
        'z') || (self.0[i] >= 'N' && self.0[i] < 'Z')
        {
            new_string.push((self.0[i] as u8 - 13) as
            char);
        } else {
            new_string.push(self.0[i]);
        }
    }
    new_string
}
```

11. Let's try compiling that program. You will quickly find it is not working, with all errors being `String` cannot be indexed by `usize`. Remember that Rust handles Unicode by default? Indexing a string will create all sorts of complications, as Unicode characters have different sizes: some are 1 byte but others can be 2, 3, or 4 bytes. With regard to index, what exactly are we saying? Is index means the byte position in a String, grapheme, or Unicode scalar values?

In the Rust language, we have primitive types such as u8, char, fn, str, and many more. In addition to those primitive types, Rust also defines a lot of modules in the standard library, such as string, io, os, fmt, and thread. These modules contain many building blocks for programming. For example, the std::string::String struct deals with String. Important programming concepts such as comparison and iteration are also defined in these modules, for example, std::cmp::Eq to compare an instance of a type with another instance. The Rust language also has std::iter::Iterator to make a type iterable. Fortunately, for String, we already have a method to do iteration.

12. Let's modify our code a little bit:

```
fn encrypt(&self) -> String {
    let mut new_string = String::new();
    for ch in self.0.chars() {
        if (ch >= 'a' && ch < 'n') || (ch >= 'A' &&
        ch < 'N') {
            new_string.push((ch as u8 + 13) as char);
        } else if (ch >= 'n' && ch < 'z') || (ch >=
        'N' && ch < 'Z') {
            new_string.push((ch as u8 - 13) as char);
        } else {
            new_string.push(ch);
        }
    }
    new_string
}
```

13. There are two ways of returning; the first one is using the return keyword such as return new_string;, or we can write just the variable without a semicolon in the last line of a function. You will see that it's more common to use the second form.

14. The preceding code works just fine, but we can make it more idiomatic. First, let's process the iterator without the for loop. Let's remove the new string initialization and use the map() method instead. Any type implementing std::iter::Iterator will have a map() method that accepts a closure as the parameter and returns std::iter::Map. We can then use the collect() method to collect the result of the closure into its own String:

```
fn encrypt(&self) -> Result<String, Box<dyn Error>> {
    self.0
```

```
        .chars()
        .map(|ch| {
            if (ch >= 'a' && ch < 'n') || (ch >= 'A'
            && ch < 'N') {
                (ch as u8 + 13) as char
            } else if (ch >= 'n' && ch < 'z') || (
            ch >= 'N' && ch < 'Z') {
                (ch as u8 - 13) as char
            } else {
                ch
            }
        })
        .collect()
}
```

The map() method accepts a closure in the form of |x| . . .. We then use the captured individual items that we get from chars() and process them.

If you look at the closure, you'll see we don't use the return keyword either. If we don't put the semicolon in a branch and it's the last item, it will be considered as a return value.

Using the if block is good, but we can also make it more idiomatic. One of the Rust language's strengths is the powerful match control flow.

15. Let's change the code again:

```
fn encrypt(&self) -> String {
    self.0
        .chars()
        .map(|ch| match ch {
            'a'..='m' | 'A'..='M' => (ch as u8 + 13)
            as char,
            'n'..='z' | 'N'..='Z' => (ch as u8 - 13)
            as char,
            _ => ch,
        })
        .collect()
}
```

That looks a lot cleaner. The pipe (|) operator is a separator to match items in an arm. The Rust matcher is exhaustive, which means that the compiler will check whether all possible values of the matcher are included in the matcher or not. In this case, it means all characters in Unicode. Try removing the last arm and compiling it to see what happens if you don't include an item in a collection.

You can define a range by using .. or ..=. The former means we are excluding the last element, and the latter means we are including the last element.

16. Now that we have implemented our simple encryptor, let's use it in our main application:

```
fn main() {
    ...
    io::stdin()
    .read_line(&mut user_input)
    .expect("Cannot read input");

    println!(
        "Your encrypted string: {}",
        encryptor::rot13::Rot13(user_input).encrypt()
    );
}
```

Right now, when we try to compile it, the compiler will show an error. Basically, the compiler is saying you cannot use a trait function if the trait is not in the scope, and the help from the compiler is showing what we need to do.

17. Put the following line above the main() function and the compiler should produce a binary without any error:

```
use encryptor::Encryptable;
```

18. Let's try running the executable:

```
> ./main
Input the string you want to encrypt:
asdf123
Your encrypted string: nfqs123

> ./main
Input the string you want to encrypt:
```

```
nfqs123
Your encrypted string: asdf123
```

We have finished our program and we improved it with real-world encryption. In the next section, we're going to learn how to search for and use third-party libraries and incorporate them into our application.

Packages and Cargo

Now that we know how to create a simple program in Rust, let's explore **Cargo**, the Rust package manager. Cargo is a **command-line application** that manages your application dependencies and compiles your code.

Rust has a community package registry at `https://crates.io`. You can use that website to search for a library that you can use in your application. Don't forget to check the license of the library or application that you want to use. If you register on that website, you can use Cargo to publicly distribute your library or binary.

How do we install Cargo into our system? The good news is Cargo is already installed if you install the Rust toolchain in the stable channel using `rustup`.

Cargo package layout

Let's try using Cargo in our application. First, let's copy the application that we wrote earlier:

```
cp -r 02ComplexProgram  03Packages
cd 03Packages
cargo init . --name our_package
```

Since we already have an existing application, we can initialize our existing application with `cargo init`. Notice we add the `--name` option because we are prefixing our folder name with a number, and a Rust package name cannot start with a number.

If we are creating a new application, we can use the `cargo new package_name` command. To create a library-only package instead of a binary package, you can pass the `--lib` option to `cargo new`.

You will see two new files, `Cargo.toml` and `Cargo.lock`, inside the folder. The `.toml` file is a file format commonly used as a configuration file. The `lock` file is generated automatically by Cargo, and we don't usually change the content manually. It's also common to add `Cargo.lock` to your source code versioning application ignore list, such as `.gitignore`, for example.

Let's check the content of the `Cargo.toml` file:

```
[package]
name = "our_package"
version = "0.1.0"
edition = "2021"

# See more keys and their definitions at
https://doc.rust-lang.org/cargo/reference/manifest.html

[dependencies]

[[bin]]
name = "our_package"
path = "main.rs"
```

As you can see, we can define basic things for our application such as `name` and `version`. We can also add important information such as authors, homepage, repository, and much more. We can also add dependencies that we want to use in the Cargo application.

One thing that stands out is the edition configuration. The Rust edition is an optional marker to group various Rust language releases that have the same compatibility. When Rust 1.0 was released, the compiler did not have the capability to know the `async` and `await` keywords. After `async` and `await` were added, it created all sorts of problems with older compilers. The solution to that problem was to introduce Rust editions. Three editions have been defined: 2015, 2018, and 2021.

Right now, the Rust compiler can compile our package perfectly fine, but it is not very idiomatic because a Cargo project has conventions on file and folder names and structures. Let's change the files and directory structure a little bit:

1. A package is expected to reside in the src directory. Let's change the Cargo.toml file [[bin]] path from "main.rs" to "src/main.rs".

2. Create the src directory inside our application folder. Then, move the main.rs file and the encryptor folder to the src folder.

3. Add these lines to Cargo.toml after [[bin]]:

    ```
    [lib]
    name = "our_package"
    path = "src/lib.rs"
    ```

4. Let's create the src/lib.rs file and move this line from src/main.rs to src/lib.rs:

    ```
    pub mod encryptor;
    ```

5. We can then simplify using both the rot13 and Encryptable modules in our main.rs file:

    ```
    use our_package::encryptor::{rot13, Encryptable};
    use std::io;

    fn main() {
        ...

        println!(
            "Your encrypted string: {}",
            rot13::Rot13(user_input).encrypt()
        );
    }
    ```

6. We can check whether there's an error that prevents the code from being compiled by typing `cargo check` in the command line. It should produce something like this:

```
> cargo check
    Checking our_package v0.1.0
    (/Users/karuna/Chapter01/03Packages)
    Finished dev [unoptimized + debuginfo] target(s)
    in 1.01s
```

7. After that, we can build the binary using the `cargo build` command. Since we didn't specify any option in our command, the default binary should be unoptimized and contain debugging symbols. The default location for the generated binary is in the `target` folder at the root of the workspace:

```
$ cargo build
   Compiling our_package v0.1.0
   (/Users/karuna/Chapter01/03Packages)
    Finished dev [unoptimized + debuginfo] target(s)
    in 5.09s
```

You can then run the binary in the `target` folder as follows:

```
./target/debug/our_package
```

debug is enabled by the default dev profile, and `our_package` is the name that we specify in `Cargo.toml`.

If you want to create a release binary, you can specify the `--release` option, `cargo build --release`. You can find the release binary in `./target/release/our_package`.

You can also type `cargo run`, which will compile and run the application for you.

Now that we have arranged our application structure, let's add real-world encryption to our application by using a third-party crate.

Using third-party crates

Before we implement another encryptor using a third-party module, let's modify our application a little bit. Copy the previous 03Packages folder to the new folder, 04Crates, and use the folder for the following steps:

1. We will rename our Encryptor trait as a Cipher trait and modify the functions. The reason is that we only need to think about the output of the type, not the encrypt process itself:

    - Let's change the content of src/lib.rs to pub mod cipher;.

    - After that, rename the encryptor folder as cipher.

    - Then, modify the Encryptable trait into the following:

    ```
    pub trait Cipher {
        fn original_string(&self) -> String;
        fn encrypted_string(&self) -> String;
    }
    ```

    The reality is we only need functions to show the original string and the encrypted string. We don't need to expose the encryption in the type itself.

2. After that, let's also change src/cipher/rot13.rs to use the renamed trait:

    ```
    impl super::Cipher for Rot13 {
        fn original_string(&self) -> String {
            String::from(&self.0)
        }
        fn encrypted_string(&self) -> String {
            self.0
                .chars()
                .map(|ch| match ch {
                    'a'..='m' | 'A'..='M' => (ch as u8 +
                    13) as char,
                    'n'..='z' | 'N'..='Z' => (ch as u8 -
                    13) as char,
                    _ => ch,
                })
                .collect()
    ```

```
        }
    }
```

3. Let's also modify `main.rs` to use the new trait and function:

    ```
    use our_package::cipher::{rot13, Cipher};

    ...

    fn main() {

        ...

        println!(
            "Your encrypted string: {}",
            rot13::Rot13(user_input).encrypted_string()
        );
    }
    ```

    The next step is to determine what encryption and library we want to use for our new type. We can go to `https://crates.io` and search for an available crate. After searching for a real-world encryption algorithm on the website, we found `https://crates.io/crates/rsa`. We found that the RSA algorithm is a secure algorithm, the crate has good documentation and has been audited by security researchers, the license is compatible with what we need, and there's a huge number of downloads. Aside from checking the source code of this library, all indications show that this is a good crate to use. Luckily, there's an install section on the right side of that page. Besides the `rsa` crate, we are also going to use the `rand` crate, since the RSA algorithm requires a random number generator. Since the generated encryption is in bytes, we must encode it somehow to `string`. One of the common ways is to use `base64`.

4. Add these lines in our `Cargo.toml` file, under the [dependencies] section:

    ```
    rsa = "0.5.0"
    rand = "0.8.4"
    base64 = "0.13.0"
    ```

5. The next step should be adding a new module and typing using the rsa crate. But, for this type, we want to modify it a little bit. First, we want to create an **associated function**, which might be called a constructor in other languages. We want to then encrypt the input string in this function and store the encrypted string in a field. There's a saying that all data not in processing should be encrypted by default, but the fact is that we as programmers rarely do this.

    Since RSA encryption is dealing with byte manipulation, there's a possibility of errors, so the return value of the associated function should be wrapped in the Result type. There's no compiler rule, but if a function cannot fail, the return should be straightforward. Regardless of whether or not a function can produce a result, the return value should be Option, but if a function can produce an error, it's better to use Result.

    The encrypted_string() method should return the stored encrypted string, and the original_string() method should decrypt the stored string and return the plain text.

    In src/cipher/mod.rs, change the code to the following:

    ```
    pub trait Cipher {
        fn original_string(&self) -> Result<String,
        Box<dyn Error>>;
        fn encrypted_string(&self) -> Result<String,
        Box<dyn Error>>;
    }
    ```

6. Since we changed the definition of the trait, we have to change the code in src/cipher/rot13.rs as well. Change the code to the following:

    ```
    use std::error::Error;


    pub struct Rot13(pub String);


    impl super::Cipher for Rot13 {
        fn original_string(&self) -> Result<String,
        Box<dyn Error>> {
            Ok(String::from(&self.0))
        }


        fn encrypted_string(&self) -> Result<String,
    ```

```
Box<dyn Error>> {
    Ok(self
        .0
        ...
        .collect())
    }
}
```

7. Let's add the following line in the `src/cipher/mod.rs` file:

    ```
    pub mod rsa;
    ```

8. After that, create `rsa.rs` inside the `cipher` folder and create the `Rsa` struct inside it. Notice that we use `Rsa` instead of `RSA` as the type name. The convention is to use `CamelCase` for type:

    ```
    use std::error::Error;

    pub struct Rsa {
        data: String,
    }

    impl Rsa {
        pub fn new(input: String) -> Result<Self, Box<
        dyn Error>> {
            unimplemented!();
        }
    }

    impl super::Cipher for Rsa {
        fn original_string(&self) -> Result<String, ()> {
            unimplemented!();
        }

        fn encrypted_string(&self) -> Result<String, ()> {
            Ok(String::from(&self.data))
        }
    }
    ```

There are a couple of things we can observe. The first one is the data field does not have the pub keyword since we want to make it private. You can see that we have two impl blocks: one is for defining the methods of the Rsa type itself, and the other is for implementing the Cipher trait.

Also, the new() function does not have self, mut self, &self, or &mut self as the first parameter. Consider it as a static method in other languages. This method is returning Result, which is either Ok(Self) or Box<dyn Error>. The Self instance is the instance of the Rsa struct, but we'll discuss Box<dyn Error> later when we talk about error handling in *Chapter 7, Handling Errors in Rust and Rocket*. Right now, we haven't implemented this method, hence the usage of the unimplemented!() macro. Macros in Rust look like a function but with an extra bang (!).

9.   Now, let's implement the associated function. Modify src/cipher/rsa.rs:

```rust
use rand::rngs::OsRng;
use rsa::{PaddingScheme, PublicKey, RsaPrivateKey};
use std::error::Error;


const KEY_SIZE: usize = 2048;


pub struct Rsa {
    data: String,
    private_key: RsaPrivateKey,
}


impl Rsa {
    pub fn new(input: String) -> Result<Self, Box<
dyn Error>> {
        let mut rng = OsRng;
        let private_key = RsaPrivateKey::new(&mut rng,
        KEY_SIZE)?;
        let public_key = private_key.to_public_key();
        let input_bytes = input.as_bytes();
        let encrypted_data =
            public_key.encrypt(&mut rng, PaddingScheme
            ::new_pkcs1v15_encrypt(), input_bytes)?;
        let encoded_data =
```

```
            base64::encode(encrypted_data);
            Ok(Self {
                data: encoded_data,
                private_key,
            })
        }
    }
```

The first thing we do is declare the various types we are going to use. After that, we define a constant to denote what size key we are going to use.

If you understand the RSA algorithm, you already know that it's an asymmetric algorithm, meaning we have two keys: a public key and a private key. We use the public key to encrypt data and use the private key to decrypt the data. We can generate and give the public key to the other party, but we don't want to give the private key to the other party. That means we must store the private key inside the struct as well.

The `new()` implementation is pretty straightforward. The first thing we do is declare a random number generator, `rng`. We then generate the RSA private key. But, pay attention to the question mark operator (?) on the initialization of the private key. If a function returns `Result`, we can quickly return the error generated by calling any method or function inside it by using (?) after that function.

Then, we generate the RSA public key from a private key, encode the input string as bytes, and encrypt the data. Since encrypting the data might have resulted in an error, we use the question mark operator again. We then encode the encrypted bytes as a `base64` string and initialize `Self`, which means the `Rsa` struct itself.

10. Now, let's implement the `original_string()` method. We should do the opposite of what we do when we create the struct:

```
fn original_string(&self) -> Result<String, Box<dyn
Error>> {
    let decoded_data = base64::decode(&self.data)?;
    let decrypted_data = self
        .private_key
        .decrypt(PaddingScheme::
        new_pkcs1v15_encrypt(), &decoded_data)?;
    Ok(String::from_utf8(decrypted_data)?)
}
```

First, we decode the `base64` encoded string in the `data` field. Then, we decrypt the decoded bytes and convert them back to a string.

11. Now that we have finished our `Rsa` type, let's use it in our `main.rs` file:

```rust
fn main() {
    ...
    println!(
        "Your encrypted string: {}",
        rot13::Rot13(user_input).encrypted_
        string().unwrap()
    );

    println!("Input the string you want to encrypt:");

    let mut user_input = String::new();

    io::stdin()
        .read_line(&mut user_input)
        .expect("Cannot read input");

    let encrypted_input = rsa::Rsa::new(
    user_input).expect("");
    let encrypted_string = encrypted_input.encrypted_
    string().expect("");

    println!("Your encrypted string: {}",
    encrypted_string);

    let decrypted_string = encrypted_input
    .original_string().expect("");
    println!("Your original string: {}",
    decrypted_string);
}
```

Some of you might wonder why we redeclared the `user_input` variable. The simple explanation is that Rust already moved the resource to the new `Rot13` type, and Rust does not allow the reuse of the moved value. You can try commenting on the second variable declaration and compile the application to see the explanation. We will discuss the Rust borrow checker and moving in more detail in *Chapter 9, Displaying Users' Post*.

Now, try running the program by typing `cargo run`:

```
$ cargo run
   Compiling cfg-if v1.0.0
   Compiling subtle v2.4.1
   Compiling const-oid v0.6.0
   Compiling ppv-lite86 v0.2.10
   ...
   Compiling our_package v0.1.0
   (/Users/karuna//Chapter01/04Crates)
    Finished dev [unoptimized + debuginfo] target(s)
    in 3.17s
     Running `target/debug/our_package`
Input the string you want to encrypt:
first
Your encrypted string: svefg

Input the string you want to encrypt:
second
Your encrypted string: lhhb9RvG9zI75U2VC3FxvfUujw0cVqqZFg-
PXhNixQTF7RoVBEJh2inn7sEefDB7eNlQcf09lD2nULfgc2mK55ZE+UU-
cYzbMDu45oTaPiDPog4L6FRVpbQR27bkOj9Bq1KS+QAvRtxtTbTa1L5/
OigZbqBc2QOm2yHLCimMPeZKhLBtK2whhtzIDM815AYTBg+rA688ZfB-
7ZI4FSRm4/h22kNzSPo1DECI04ZBprAq4hWHxEKRwtn5TkRLhClGFL-
SYKkY7Ajjr3EOf4QfkUvFFhZ0qRDndPI5c9RecavofVLxECrYfv5ygYR-
mW3B1cJn4vcBhVKfQF0JQ+vs+FuTUpw==
Your original string: second
```

You will see that Cargo automatically downloaded the dependencies and builds them one by one. Also, you might notice that encrypting using the Rsa type took a while. Isn't Rust supposed to be a fast system language? The RSA algorithm itself is a slow algorithm, but that's not the real cause of the slowness. Because we are running the program in a development profile, the Rust compiler generates an application binary with all the debugging information and does not optimize the resulting binary. On the other hand, if you build the application using the --release flag, the compiler generates an optimized application binary and strips the debugging symbols. The resulting binary compiled with the release flag should execute faster than the debug binary. Try doing it yourself so you'll remember how to build a release binary.

In this section, we have learned about Cargo and third-party packages, so next, let's find out where to find help and documentation for the tools that we have used.

Tools and getting help

Now that we have created a pretty simple application, you might be wondering what tools we can use for development, and how to find out more about Rust and get help.

Tools

Besides Cargo, there are a couple more tools we can use for Rust application development:

- **rustfmt**

  This program is for formatting your source code so it follows the Rust style guide. You can install it by using rustup (rustup component add rustfmt). Then, you can integrate it with your favorite text editor or use it from the command line. You can read more about rustfmt at https://github.com/rust-lang/rustfmt.

- **clippy**

  Does the name remind you of something? clippy is useful for linting your Cargo application using various lint rules. Right now, there are more than 450 lint rules you can use. You can install it using this command: rustup component add clippy. Afterward, you can use it in the Cargo application by running cargo clippy. Can you try it in the Cargo application that we wrote earlier? You can read more about clippy at https://github.com/rust-lang/rust-clippy.

Text editor

Most likely, the text editor of your choice already supports the Rust language, or at least syntax highlighting Rust. You can install the Rust language server if you want to add important functionalities such as go to definition, go to implementation, symbol search, and code completion. Most popular text editors already support the language server, so you can just install an extension or other integration method to your text editor:

- **The Rust language server**

  You can install it using the `rustup` command: `rustup component add rls rust-analysis rust-src`. Then, you can integrate it into your text editor. For example, if you are using **Visual Studio Code**, you can install the Rust extension and enable `rls`.

  You can read more about it at `https://github.com/rust-lang/rls`.

- **Rust analyzer**

  This application is poised to be the Rust language server 2.0. It's still considered to be in alpha as of the writing of this book, but in my experience, this application works well with regular updates. You can find the executable for this one at `https://github.com/rust-analyzer/rust-analyzer/releases`, and then configure your editor language server to use this application. You can read more about it at `https://rust-analyzer.github.io`.

Getting help and documentation

There are a few important documents that you might want to read to find help or references:

- **The Rust programming language book**: This is the book that you want to read if you want to understand more about the Rust programming language. You can find it online at `https://doc.rust-lang.org/book/`.

- **Rust by Example**: This documentation is a collection of small examples that show the concepts of the Rust language and its standard library's capabilities. You can read it online at `https://doc.rust-lang.org/rust-by-example/index.html`.

- **Standard library documentation**: As a programmer, you will refer to this standard library documentation. You can read more about standard libraries, their modules, the function signatures, what standard libraries' functions do, read the examples, and more. Find it at `https://doc.rust-lang.org/std/index.html`.

- **The Cargo book**: If you are interested in Cargo and related information such as the `Cargo.toml` manifest format, you can read more about it at `https://doc.rust-lang.org/cargo/index.html`.

- **Rust style guidelines**: The Rust language, like other programming languages, has style guidelines. These guidelines tell a programmer what the convention for naming is, about whitespaces, how to use constants, and other idiomatic conventions for a Rust program. Read more about it at `https://doc.rust-lang.org/1.0.0/style/`.

- **Docs.rs**: Suppose you are using a third-party crate, such as the `rsa` crate that we used earlier. To find documentation for that library, you can go to `https://crates.io` and search for the crate's page, then go to the right pane and go to the documentation section. Or, you can go to `https://docs.rs` and search for the crate name and find the documentation for it.

- **Rustup doc**: This documentation is not online, but you can install it using `rustup` (`rustup component add rust-docs`). Then, you can open documentation in your browser while offline using the `rustup doc` command. If you want to open standard library documentation offline, you can type `rustup doc --std`. There are other documents you can open; try and see what they are by using `rustup doc --help`.

- **The Rust user forum**: If you want to get help or help other Rust programmers, you can find it all over the internet. There's a dedicated forum to discuss Rust-related topics at `https://users.rust-lang.org/`.

Summary

In this chapter, we had a brief overview of the Rust language. We learned about the Rust toolchain and how to install it as well as the tools required for Rust development. After that, we created two simple programs, used Cargo, and imported third-party modules to improve our program. Now that you can write a small program in the Rust language, explore! Try creating more programs or experimenting with the language. You can try *Rust by Example* to see what features we can use in our programs. In subsequent chapters, we will learn more about Rocket, a web framework written in the Rust language.

2
Building Our First Rocket Web Application

In this chapter, we're going to explore Rocket, a framework to create web applications using the Rust programming language. We will learn a little bit about Rocket before we create our first web application using the Rocket framework. After that, we will learn how to configure our Rocket web application. Finally, we will explore how to get help for this web framework at the end of this chapter.

In this chapter, we're going to cover the following main topics:

- Introducing Rocket – a web framework written in the Rust language
- Creating our first Rocket web application
- Configuring our Rocket web application
- Getting help

Technical requirements

For this and subsequent chapters, you will need to have the requirements mentioned in *Chapter 1, Introducing the Rust Language*, and the Rust toolchain installed. If you still don't have the Rust compiler toolchain installed, please follow the installation guide in *Chapter 1, Introducing the Rust Language*. Also, it would be helpful to have a text editor with a Rust extension installed and Rust tools such as `rustfmt` or `clippy`. If you don't have a text editor installed already, you can use open source software such as Visual Studio Code with the `rust-analyzer` extension. As we're going to make HTTP requests to the application that we're going to create, you should have a web browser or other HTTP client installed.

Finally, the Rocket framework has a few releases, all of which are slightly different. We will only be discussing Rocket *0.5.0*. Don't worry if you are planning to use a different version of Rocket, as the terminology and concepts are almost the same. Use the API documentation mentioned in the *Getting help* section of this chapter to see the correct documentation for your Rocket framework version.

The code for this chapter can be found at `https://github.com/PacktPublishing/Rust-Web-Development-with-Rocket/tree/main/Chapter02`.

Introducing Rocket – a web framework written in the Rust language

The development of the Rocket web framework began as a project of Sergio Benitez in 2016. For a long time, it was being created with a lot of Rust macrosystems to simplify the development process; because of this, a stable Rust compiler could not be used until recently, in 2021. During the development process, async/await capabilities were added to Rust. Rocket began to incorporate async/await until the issue tracker for it closed in 2021.

Rocket is a fairly simple web framework without many bells and whistles, such as database **Object-relational mapping** (**ORM**) or mailing systems. Programmers can extend Rocket's capabilities using other Rust crates, for example, by adding third-party logging or connecting to memory store applications.

The HTTP request life cycle in Rocket

Handling HTTP requests is an integral part of web applications. The Rocket web framework treats incoming HTTP requests as a life cycle. The first thing the web application does is checks and determines which function or functions will be able to handle the incoming request. This part is called **routing**. For example, if there's a *GET/ something* incoming request, Rocket will check all the registered routes for matches.

After routing, Rocket will perform the **validation** of the incoming request against types and guards declared in the first function. If the result does not match and the next route handling function is available, Rocket will continue validation against the next function until there are no more functions available to handle that incoming request.

After validation, Rocket will then **process** with what the programmer wrote in the body of the function. For example, a programmer creates a SQL query with the data from the request, sends the query to the database, retrieves the result from the database, and creates an HTML using the result.

Rocket will finally return a **response**, which contains the HTTP status, headers, and body. The request life cycle is then complete.

To recap, the life cycle of a Rocket request is **Routing → Validation → Processing → Response**. Next, let's discuss how the Rocket application starts.

Rocket launch sequence

Like a real-life rocket, we start by building the Rocket application. In the building process, we mount the **routes** (functions that handle incoming requests) to the Rocket application.

In the building process, the Rocket application also manages various **states**. A state is a Rust object that can be accessed in a route handler. For example, let's say we want a logger that sends events to a logging server. We initialize the logger object when we build the Rocket application, and when there's an incoming request, we can use the already managed logger object in the request handler.

Still in the building process, we can also attach **fairings**. In many web frameworks, there is usually a middleware component that filters, inspects, authorizes, or modifies incoming HTTP requests or responses. In the Rocket framework, the function that provides middleware functionality is called a fairing. For example, if we want to have a **universally unique identifier** (UUID) for every HTTP request for audit purposes, we first create a fairing that generates a random UUID and appends it to the request HTTP header. We also make the fairing append the same UUID to the generated HTTP response. Next, we attach it to Rocket. This fairing will then intercept the incoming request and response and modify it.

After the Rocket application is built and ready, the next step is, of course, launching it. Yay, the launch is successful! Now, our Rocket application is operational and ready to serve incoming requests.

Now that we have an overview of the Rocket application, let's try creating a simple Rocket application.

Creating our first Rocket web application

In this section, we are going to create a very simple web application that handles only one HTTP path. Follow these steps to create our first Rocket web application:

1. Create a new Rust application using Cargo:

    ```
    cargo new 01hello_rocket --name hello_rocket
    ```

    We are creating an application named `hello_rocket` in a folder named `01hello_rocket`.

2. After that, let's modify the `Cargo.toml` file. Add the following line after `[dependencies]`:

    ```
    rocket = "0.5.0-rc.1"
    ```

3. Append the following lines at the top of the `src/main.rs` file:

    ```
    #[macro_use]
    extern crate rocket;
    ```

    Here, we are telling the Rust compiler to use macros from the Rocket crate by using the `#[macro_use]` attribute. We can skip using that attribute, but that would mean we must specify `use` for every single macro that we are going to use.

4. Add the following line to tell the compiler that we are using the definition from the Rocket crate:

    ```
    use rocket::{Build, Rocket};
    ```

5. After that, let's create our first HTTP handler. Add the following lines after the preceding ones:

    ```
    #[get("/")]
    fn index() -> &'static str {
        "Hello, Rocket!"
    }
    ```

Here, we define a function that returns a reference to a `str`. The `'a` in the Rust language means that a variable has a `'a` lifetime. The life span of a reference depends on many things. We will discuss these in *Chapter 9, Displaying User's Post*, when we discuss object scope and lifetime in more depth. But, a `'static` notation is special because it means that it will last as long as the application is still alive. We can also see that the return value is `"Hello, Rocket"` since it is the last line and we did not put a semicolon at the end.

But, what is the `#[get("/")]` attribute? Remember before when we used the `#[macro_use]` attribute? The `rocket::get` attribute is a macro attribute that specifies the HTTP method a function handles, and the route, HTTP paths, and parameters it handles. There are seven method-specific route attributes that we can use: `get`, `put`, `post`, `delete`, `head`, `options`, and `patch`. All of them correspond to their respective HTTP method name.

We can also use alternate macros to specify route handlers by replacing the attribute macro with the following:

```
#[route(GET, path = "/")]
```

6. Next, delete the `fn main()` function and add the following lines:

```
#[launch]
fn rocket() -> Rocket<Build> {
    rocket::build().mount("/", routes![index])
}
```

We have created a function that will generate the `main` function because we used the `#[launch]` attribute. Inside the function, we built the Rocket and mounted the routes that have an `index` function to the `"/"` path.

7. Let's try running the `hello_rocket` application:

```
> cargo run
    Updating crates.io index
    Compiling proc-macro2 v1.0.28
...
    Compiling hello_rocket v0.1.0 (/Users/karuna/
    Chapter02/01hello_rocket)
```

```
      Finished dev [unoptimized + debuginfo] target(s)
      in 1m 39s
        Running `target/debug/hello_rocket`
```

🔧 Configured for debug.
    >> address: 127.0.0.1
    >> port: 8000
    >> workers: 8
    >> ident: Rocket
    >> keep-alive: 5s
    >> limits: bytes = 8KiB, data-form = 2MiB, file =
    1MiB, form = 32KiB, json = 1MiB, msgpack = 1MiB,
    string = 8KiB
    >> tls: disabled
    >> temp dir: /var/folders/gh/
    kgsn28fn3hvflpcfq70x6f1w0000gp/T/
    >> log level: normal
    >> cli colors: true
    >> shutdown: ctrlc = true, force = true, signals =
    [SIGTERM], grace = 2s, mercy = 3s

🐛 Routes:
    >> (index) GET /

🪁 Fairings:
    >> Shield (liftoff, response, singleton)

🛡 Shield:
    >> X-Content-Type-Options: nosniff
    >> X-Frame-Options: SAMEORIGIN
    >> Permissions-Policy: interest-cohort=()

🚀 Rocket has launched from http://127.0.0.1:8000

You can see the application printed the application configuration, such as keep-alive timeout duration, request size limits, log level, routes, and many more that you typically see in an HTTP server to the terminal. After that, the application printed the various parts of the Rocket. We created a single-route index function, which handles GET /.

Then, there is the default built-in fairing, **Shield**. Shield works by injecting HTTP security and privacy headers to all responses by default.

We also see that the application was successfully launched and is now accepting requests on the address `127.0.0.1` and port `8000`.

8. Now, let's test whether the application is really accepting requests. You can use a web browser or any HTTP client since it's a very simple request, but if you use the command line, don't stop the running application; open another terminal:

```
> curl http://127.0.0.1:8000
Hello, Rocket!
```

You can see that the application responded perfectly. You can also see the log of the application:

```
GET /:
    >> Matched: (index) GET /
    >> Outcome: Success
    >> Response succeeded.
```

9. Now, let's try requesting something that does not exist in our application:

```
> curl http://127.0.0.1:8000/somepath
<!DOCTYPE html>
<html lang="en">
<head>
    <meta charset="utf-8">
    <title>404 Not Found</title>
</head>
<body align="center">
    ...
</body>
</html>
```

Now, let's see what's going on in the application terminal output:

```
GET /somepath:
    >> No matching routes for GET /somepath.
    >> No 404 catcher registered. Using Rocket default.
    >> Response succeeded.
```

We now know that Rocket already has default handlers for `404` status situations.

Let's recall the Rocket life cycle, *Routing → Validation → Processing → Response*. The first request to `http://127.0.0.1:8000/` was successful because the application found the handler for the `GET` `/` route. Since we didn't create any validation in the application, the function then performed some very simple processing, returning a string. The Rocket framework already implemented the `Responder` trait for `&str`, so it created and returned an appropriate HTTP response. The other request to `/somepath` did not pass the routing part, and we did not create any error handler, so the Rocket application returned a default error handler for this request.

Try opening it in the browser and inspecting the response using developer tools, or try running the `curl` command again in verbose mode to see the complete HTTP response, `curl -v http://127.0.0.1:8000/` and `curl -v http://127.0.0.1:8000/somepath`:

```
$ curl -v http://127.0.0.1:8000/somepath
*    Trying 127.0.0.1...
* TCP_NODELAY set
* Connected to 127.0.0.1 (127.0.0.1) port 8000 (#0)
> GET /somepath HTTP/1.1
> Host: 127.0.0.1:8000
> User-Agent: curl/7.64.1
> Accept: */*
>
< HTTP/1.1 404 Not Found
< content-type: text/html; charset=utf-8
< server: Rocket
< x-frame-options: SAMEORIGIN
< x-content-type-options: nosniff
< permissions-policy: interest-cohort=()
< content-length: 383
< date: Fri, 17 Aug 1945 03:00:00 GMT
<
<!DOCTYPE html>
...
* Connection #0 to host 127.0.0.1 left intact
</html>* Closing connection 0
```

You can see that access to / worked perfectly and access to /somepath returned an HTTP response with a 404 status and with some HTML content. There are some default privacy and security HTTP headers too, which were injected by the Shield fairing.

Congratulations! You just created your first Rocket-powered web application. What you just built is a regular Rocket web application. Next, let's modify it to be an asynchronous web application.

An asynchronous application

What is an asynchronous application? Let's say our web application is sending a query to a database. While waiting for the response for a few milliseconds, our application thread is just doing nothing. For a single user and a single request, this is not a problem. An asynchronous application is an application that allows a processor to do other tasks while there are blocking tasks, such as waiting for the response from the database. We will discuss this in detail later; right now, we just want to convert our application into an asynchronous application.

Let's modify the application that we created earlier and make it asynchronous. You can find the example folder in 02hello_rocket_async:

1. Remove the use rocket::{Build, Rocket}; line, since we are not going to use it.

2. After that, let's add the async keyword before fn index().

3. Replace the #[launch] attribute with #[rocket::main]. This is to signify that this function is going to be the main function in our application.

4. Add the async keyword and rename fn launch() to fn main().

5. We also don't want the main function to return anything, so use remove -> Rocket<build>.

6. Add .launch().await; after calling mount.

The final code should look like this:

```
#[macro_use]
extern crate rocket;

#[get("/")]
async fn index() -> &'static str {
    "Hello, Rocket!"
}
```

```
#[rocket::main]
async fn main() {
    rocket::build().mount("/", routes![
    index]).launch().await;
}
```

Stop the old version from running on the server by using the *Ctrl + C* command. You should see something like this:

```
Warning: Received SIGINT. Requesting shutdown.
Received shutdown request. Waiting for pending I/O...
```

We don't have any blocking task in our "/" handler right now, so we will not see any noticeable benefit. Now that we have created our application, let's configure it in the next section.

Configuring our Rocket web application

Let's learn how to configure our Rocket web application, starting with different profiles for different situations. Then, we will use the Rocket.toml file to configure it. And finally, we will learn how to use environment variables to configure our application.

Starting the Rocket application in different profiles

Let's run our synchronous application server without a release flag, and in another terminal, let's see whether we can benchmark it:

1. First, let's install the application using cargo install benchrs. That's right, you can install the application using Cargo too! There are very good Rust programs that you can use in your terminal, for example, ripgrep, which is one of the fastest applications for grepping string in your code.

    If you want to call the Cargo-installed application, you can use the full path or add it to your terminal path if you are using a Unix-based terminal. Append the following line to your ~/.profile or any other profile file that will be loaded by your terminal:

    ```
    export PATH="$HOME/.cargo/bin:$PATH"
    ```

    Rustup should already have added Cargo's bin folder to your path if you are using Windows.

2. Run the benchmark against your running application:

```
benchrs -c 30 -n 3000 -k http://127.0.0.1:8000/
07:59:04.934 [INFO] benchrs:0.1.8
07:59:04.934 [INFO] Spawning 8 threads
07:59:05.699 [INFO] Ran in 0.7199552s 30 connections,
3000 requests with avg request time: 6.5126667ms, median:
6ms, 95th percentile: 11ms and 99th percentile: 14ms
```

You can see that our application handles around 3,000 requests in 0.7199552 seconds. Not a bad value for a simple application if we compare it to other heavy frameworks. After that, stop the application for now.

3. Now, let's run the application again but this time in release mode. Do you still remember how to do it from the previous chapter?

```
> cargo run -release
```

It should then compile our application for release and run it.

4. After the application is ready to accept the request again, in another terminal, run the benchmark again:

```
$ benchrs -c 30 -n 3000 -k http://127.0.0.1:8000/
08:12:51.388 [INFO] benchrs:0.1.8
08:12:51.388 [INFO] Spawning 8 threads
08:12:51.513 [INFO] Ran in 0.07942524s 30 connections,
3000 requests with avg request time: 0.021333333ms,
median: 0ms,
5th percentile: 0ms and 99th percentile: 1ms
```

Again, the result is very impressive, but what's going on here? The total benchmark time now becomes roughly 0.08 seconds, almost 10 times faster than the previous total benchmark time.

To understand the reason for the speed increase, we need to know about Rocket **profiles**. A profile is a name we give for a set of configurations.

Rocket application has two meta-profiles: **default** and **global**. The default profile has all the default configuration values. If we create a profile and do not set the values of its configuration, the values from the dcfault configuration will be used. As for the global profile, if we set the configuration values, then it will override the values set in a profile.

Besides those two meta-profiles, the Rocket framework also provides two configurations. When running or compiling the application in release mode, Rocket will use the **release** profile, and while running or compiling in debug mode, Rocket will select the **debug** profile. Running the application in release mode will obviously generate an optimized executable binary, but there are other optimizations in the application itself. For example, you will see the difference in the application output. The debug profile by default shows the output on the terminal, but the release profile by default will not show any request in the terminal output.

We can create any name for a profile, for example, development, test, staging, sandbox, or production. Use any name that makes sense to your development process. For example, in a machine used for QA testing, you might want to give the profile the name testing.

To choose which profile we want to use, we can specify it in the environment variable. Use ROCKET_PROFILE=profile_name cargo run in the command line. For example, you can write ROCKET_PROFILE=profile_name cargo run –release.

Now that we know how to start the application with a certain profile, let's learn how to create a profile and configure the Rocket application.

Configuring the Rocket web application

Rocket has a way to configure web applications. The web framework uses the **figment** crate (https://crates.io/crates/figment) for configuration. There are many providers (that is, types that implement a Provider trait). Someone can make a type that reads JSON from a file and implements a Provider trait for that type. That type can then be consumed by an application that uses the figment crate as the source of the configuration.

In Rust, there's a convention to initialize a struct with a default value if it implements a standard library trait, std::default::Default. That trait is written as follows:

```
pub trait Default {
    fn default() -> Self;
}
```

For example, a struct named StructName, which implements the Default trait, will then be called StructName::default(). Rocket has a rocket::Config struct that implements the Default trait. The default value is then used to configure the application.

If you look at the source for the `rocket::Config` struct, it is written as follows:

```
pub struct Config {
    pub profile: Profile,
    pub address: IpAddr,
    pub port: u16,
    pub workers: usize,
    pub keep_alive: u32,
    pub limits: Limits,
    pub tls: Option<TlsConfig>,
    pub ident: Ident,
    pub secret_key: SecretKey,
    pub temp_dir: PathBuf,
    pub log_level: LogLevel,
    pub shutdown: Shutdown,
    pub cli_colors: bool,
    // some fields omitted
}
```

As you can see, there are fields such as `address` and `port`, which will obviously dictate how the application behaves. If you check further in the source code, you can see the `Default` trait implementation for the `Config` struct.

Rocket also has a couple of figment providers that override the default `rocket::Config` value when we use the `rocket::build()` method in our application.

The first figment provider reads from the `Rocket.toml` file, or the file that we specify if we run the application with the `ROCKET_CONFIG` environment variable. If we specify `ROCKET_CONFIG` (for example, `ROCKET_CONFIG=our_config.toml`), it will search `our_config.toml` at the root directory of the application. If the application cannot find that configuration file, then the application will look in the parent folder until it reaches the root of the filesystem. If we specify an absolute path, for example, `ROCKET_CONFIG=/some/directory/our_config.toml`, then the application will only search for the file in that location.

The second figment provider reads the value from Environment variables. We will see how to do it later, but first, let's try configuring the Rocket application using the `Rocket.toml` file.

Configuring the Rocket application using Rocket.toml

The first thing we need to know is the list of keys we can use in the configuration file. These are the keys that we can use in the configuration file:

- `address`: The application will serve at this address.

- `port`: The application will serve on this port.

- `workers`: The application will use this number of threads.

- `ident`: If we specify `false`, the application will not put an identity in the server HTTP header; if we specify `string`, the application will use it as an identity in the server HTTP header.

- `keep_alive`: Keep-alive timeout in seconds. Use 0 to disable `keep_alive`.

- `log_level`: The maximum level to log (off/normal/debug/critical).

- `temp_dir`: The path to a directory to store temporary files.

- `cli_colors`: Use colors and emojis on the log, or not. This is useful to disable bells and whistles in the release environment.

- `secret_key`: The Rocket application has a type to store private cookies in the application. The private cookies are encrypted by this key. The key length is 256 bits, and you can generate it using tools such as `openssl rand -base64 32`. Since this is an important key, you might want to keep it in a safe place.

- `tls`: Use `tls.key` and `tls.certs` to enter the path to your TLS (Transport Layer Security) key and certificate file.

- `limits`: This configuration is nested and is used to limit the server read size. You can write the value in multibyte units such as 1 MB (megabyte) or 1 MiB (mebibyte). There are several default options:

  - `limits.form` – 32 KiB

  - `limits.data-form` – 2 MiB

  - `limits.file` – 1 MiB

  - `limits.string` – 8 KiB

  - `limits.bytes` – 8 KiB

  - `limits.json` – 1 MiB

  - `limits.msgpack` – 1 MiB

- `shutdown`: If a web application is terminated abruptly when it's still processing something, the data being processed might accidentally get corrupted. For example, let's say a Rocket application is in the middle of sending updated data to the database server, but the process is then terminated suddenly. As a result, there is data inconsistency. This option configures Rocket's smooth shutdown behavior. Like `limits`, it has several subconfigurations:

  - `shutdown.ctrlc` – Does the application ignore the *Ctrl* + *C* keystrokes or not?

  - `shutdown.signals` – An array of Unix signals that trigger a shutdown. Only works on Unix or Unix-like operating systems.

  - `shutdown.grace` – The number of seconds in which to finish outstanding server I/O before stopping it.

  - `shutdown.mercy` – The number of seconds in which to finish outstanding connection I/O before stopping it.

  - `shutdown.force` – Specifies whether or not to kill a process that refuses to cooperate.

Now that we know what keys we can use, let's try configuring our application. Remember what port our application is running on? Suppose now we want to run the application on port `3000`. Let's create a `Rocket.toml` file in the root folder of our application:

```
[default]
port = 3000
```

Now, try running the application again:

```
$ cargo run
...
🚀 Rocket has launched from http://127.0.0.1:3000
```

You can see that it's working; we're running the application in port `3000`. But, what if we want to run the application in a different configuration for a different profile? Let's try adding these lines in the `Rocket.toml` file and running the application in release mode:

```
[release]
port = 9999

$ cargo run --release
...
🚀 Rocket has launched from http://127.0.0.1:9999
```

That's right, we can specify the configuration for a different profile. What do we do if our option is nested? Because this file is a `.toml` file, we can write it as follows:

```
[default.tls]
certs = "/some/directory/cert-chain.pem"
key = "/some/directory/key.pem"
```

Or, we can write it in the following way:

```
[default]
tls = { certs = "/some/directory/cert-chain.pem", key = "/some/
directory/key.pem" }
```

Now, let's see the whole file with the default configuration:

```
[default]
address = "127.0.0.1"
port = 8000
workers = 16
keep_alive = 5
ident = "Rocket"
log_level = "normal"
temp_dir = "/tmp"
cli_colors = true
## Please do not use this key, but generate your own with
`openssl rand -base64 32`
secret_key = " BCbkLMhRRtYMerGKCcboyD4Mhf6/XefvhW0Wr8Q0s1Q="

[default.limits]
form = "32KiB"
data-form = "2MiB"
file = "1MiB"
string = "8KiB"
bytes = "8KiB"
json = "1MiB"
msgpack = "1MiB"

[default.tls]
```

```
certs = "/some/directory/cert-chain.pem
key = "/some/directory/key.pem


[default.shutdown]
ctrlc = true
signals = ["term"]
grace = 5
mercy = 5
force = true
```

Even though we can create the file with the whole configuration, the best practice for using `Rocket.toml` is to rely on the default value and only write what we really need to override.

Overriding the configuration with environment variables

After checking `Rocket.toml`, the application then overrides the `rocket::Config` value again with environment variables. The application will check the availability of the `ROCKET_*` environment variables. For example, we might define `ROCKET_IDENT="Merpay"` or `ROCKET_TLS={certs="abc.pem",key="def.pem"}`. This is very useful if we are doing development and have multiple team members, or if we don't want something to exist in the configuration files and rely on environment variables, for example, when we store `secret_key` in Kubernetes Secrets. In this case, getting the `secret` value from environment variables is more secure compared to writing the value in `Rocket.toml` and committing it to your source code versioning system.

Let's try overriding the configuration by running the application with `ROCKET_PORT=4000`:

```
$ ROCKET_PORT=4000 cargo run
...
🚀 Rocket has launched from http://127.0.0.1:4000
```

The environment variable override works; we are running the application in port `4000` even though we specified port `3000` in the `Rocket.toml` file. We will learn how to extend the default `rocket::Config` with custom configuration when we configure the application to connect to a database in *Chapter 4*, *Building, Igniting, and Launching Rocket*. Now that we have learned how to configure the Rocket application, let's find out where we can get documentation and help for the Rocket web framework.

Getting help

Getting help with a web framework is essential. In this part, we will see where we can get help and documentation for the Rocket framework.

You can get help from the website of the Rocket framework itself: `https://rocket.rs/`. On that website, there is a guide as follows: `https://rocket.rs/v0.5-rc/guide/`. In the top-left corner of that page, there is a dropdown where you can choose documentation for previous versions of the Rocket web framework.

At `https://api.rocket.rs`, you can see the documentation for the API, but, unfortunately, this documentation is for the master branch of the Rocket web framework. If you want to see the API documentation for your framework version, you have to manually search for it, such as `https://api.rocket.rs/v0.3/rocket/` or `https://api.rocket.rs/v0.4/rocket/`.

There is an alternative way to generate offline documentation for the Rocket framework. Go and download the source code of Rocket from the official repository at `https://github.com/SergioBenitez/Rocket`. Then, inside the folder, type `./scripts/mk-docs.sh` to run the shell script. The generated documentation is useful because, sometimes, there are items that are different from those at `https://api.rocket.rs`. For example, the definition for `rocket::Config` and its default value in the code is a little bit different from the one in the API documentation.

Summary

In this chapter, we learned a little bit about the Rocket start sequence and request life cycle. We also created a very simple application and converted it to an asynchronous application. After that, we learned about the Rocket configuration, wrote the configuration using `Rocket.toml`, and overrode it using environment variables. Finally, we learned where to find the documentation for the Rocket framework.

Now that we have created a simple application with the Rocket web framework, let's discuss requests and responses further in the next chapter.

3
Rocket Requests and Responses

We will discuss Rocket **requests** and **responses** more in this chapter. The first section will discuss how Rocket handles incoming requests in the form of routes. We will learn about various parts of a route including HTTP methods, URIs, and paths. Then, we will create an application that uses various parts in routes. We will also talk about Rust **traits** and implement a Rust trait to create a request handler.

We are also going to discuss responses in a Rocket route handler and implement returning responses. After that, we will talk more about various built-in responder implementations and learn how to create an error handler to create a custom error when a route handler fails. Finally, we will implement a generic error handler to handle common HTTP status codes such as 404 and 500.

By the end of the chapter, you will be able to create the most important part of the Rocket framework: functions to handle incoming requests and return responses.

In this chapter, we're going to cover the following main topics:

- Understanding Rocket routes
- Implementing route handlers

- Creating responses
- Making default error handlers

Technical requirements

We still have the same technical requirements from *Chapter 2, Building Our First Rocket Web Application* for this chapter. We require a Rust compiler to be installed, along with a text editor, and an HTTP client.

You can find the source code for this chapter at `https://github.com/PacktPublishing/Rust-Web-Development-with-Rocket/tree/main/Chapter03`.

Understanding Rocket routes

We begin our chapter by discussing how Rocket handles incoming requests in the form of routes. We write functions that can be used to handle incoming requests, put route attribute above those functions, and attach the route handling functions to the Rocket. A route has an HTTP method and a **URI**, which corresponds to the URL path and URL query. The URI can be static, dynamic, or a combination of both. As well as a URI, there are other parameters in a route: **rank**, **format**, and **data**. We'll talk about them in detail later, but first, let's see how we can write a route in our code. Just like previously, let's create a new Rust application and add Rocket as a dependency. After that, let's add the following lines in the `src/main.rs` file:

```
#[macro_use]
extern crate rocket;

use rocket::{Build, Rocket};

#[derive(FromForm)]
struct Filters {
    age: u8,
    active: bool,
}

#[route(GET, uri = "/user/<uuid>", rank = 1, format = "text/plain")]
fn user(uuid: &str) { /* ... */ }
```

```
#[route(GET, uri = "/users/<grade>?<filters..>")]
fn users(grade: u8, filters: Filters) { /* ... */ }

#[launch]
fn rocket() -> Rocket<Build> {
    rocket::build().mount("/", routes![user, users])
}
```

The highlighted lines are the `route` attributes. You can only put the `route` attribute in a free function and not in a method inside `impl` of a `Struct`. Now, let's discuss the route parts in detail.

HTTP methods

The first parameter you see inside the route definition is the HTTP method. The HTTP method is defined in the `rocket::http::Method` enum. The enum has the `GET`, `PUT`, `POST`, `DELETE`, `OPTIONS`, `HEAD`, `TRACE`, `CONNECT`, and `PATCH` members, which all correspond to valid HTTP methods defined in RFCs (Request For Comments).

We can use other attributes to denote a route besides using the `#[route...]` macro. We can directly use method-specific route attributes such as `#[get...]`. There are seven method-specific route attributes: `get`, `put`, `post`, `delete`, `head`, `options`, and `patch`. We can rewrite the previous route attributes into the following lines:

```
#[get("/user/<uuid>", rank = 1, format = "text/plain")]
#[get("/users/<grade>?<filters..>")]
```

It looks simple, right? Unfortunately, we still have to use the `#[route...]` attribute if we want to handle HTTP `CONNECT` or `TRACE` as there are no method-specific route attributes for these two methods.

URI

Inside the route attribute, we can see the **URI**. The URI parameter is a string that has two parts, the **path** and the **query**. The part before the question mark (?) is the path and the part after the question mark is the query.

Both the path and the query can be divided into **segments**. The path is segmented by the slash (/), as in `/segment1/segment2`. The query is segmented by an ampersand (&), as in `?segment1&segment2`.

A segment can be **static** or **dynamic**. The static form is fixed, such as `/static` or `?static`. The dynamic segment is defined inside angle brackets (`<>`), as in `/<dynamic>` or `?<dynamic>`.

If you declare a dynamic segment, you must use the segment as a function parameter in the function following the route attribute. The following is an example of how we can use the dynamic segment by writing a new application and adding this route and function handler:

```
#[get("/<id>")]
fn process(id: u8) {/* ... */}
```

Path

The argument type in the handler function for the path must implement the `rocket::request::FromParam` trait. You might be wondering why we used `u8` as a function argument in the previous example. The answer is because Rocket has already implemented the `FromParam` trait for important types, such as `u8`. The following is a list of all types that have already implemented the `FromParam` trait:

- Primitive types such as `f32`, `f64`, `isize`, `i8`, `i16`, `i32`, `i64`, `i128`, `usize`, `u8`, `u16`, `u32`, `u64`, `u128`, and `bool`.

- Rust standard library numerical types in the `std::num` module, such as `NonZeroI8`, `NonZeroI16`, `NonZeroI32`, `NonZeroI64`, `NonZeroI128`, `NonZeroIsize`, `NonZeroU8`, `NonZeroU16`, `NonZeroU32`, `NonZeroU64`, `NonZeroU128`, and `NonZeroUsize`.

- Rust standard library net types in the `std::net` module, such as `IpAddr`, `Ipv4Addr`, `Ipv6Addr`, `SocketAddrV4`, `SocketAddrV6`, and `SocketAddr`.

- `&str` and `String`.

- `Option<T>` and `Result<T, T::Error>` where `T:FromParam`. If you are new to Rust, this syntax is for a generic type. T:FromParam means that we can use any type `T`, as long as that type implements `FromParam`. For example, we can create a `User` struct, implement `FromParam` for `User`, and use `Option<User>` as an argument in the function handler.

If you write a dynamic segment path, you must use the argument in the handler function, or else the code will fail to compile. The code will also fail if the argument type does not implement the `FromParam` trait.

Let's see the error if we don't use the argument in the handler function by removing `id: u8` from the code:

```
> cargo build
   ...
   Compiling route v0.1.0 (/Users/karuna/Chapter03/
   04UnusedParameter)
error: unused parameter
  --> src/main.rs:6:7
   |
6  | #[get("/<id>")]
   |        ^^^^^^

error: [note] expected argument named `id` here
  --> src/main.rs:7:15
   |
7  | fn process_abc() { /* ... */ }
   |               ^^
```

Then, let's write the dynamic segment, which does not implement `FromParam`. Define an empty struct and use that as an argument in the handler function:

```
struct S;

#[get("/<id>")]
fn process(id: S) { /* ... */ }
```

Again, the code will not compile:

```
> cargo build
...
   Compiling route v0.1.0 (/home/karuna/workspace/
   rocketbook/Chapter03/05NotFromParam)

error[E0277]: the trait bound `S: FromParam<'_>` is not
satisfied
  --> src/main.rs:9:16
```

```
  |
9 | fn process(id: S) { /* ... */ }
  |                     ^ the trait `FromParam<'_>` is not
implemented for `S`
  |
  = note: required by `from_param`
error: aborting due to previous error
```

We can see from the compiler output that type S must implement the FromParam trait.

There is another dynamic form in angle brackets but trailed with two full stops (..), as in /<dynamic..>. This dynamic form is called **multiple segments**.

If a regular dynamic segment must implement the FromParam trait, multiple segments must implement the rocket::request::FromSegments trait. Rocket only provides FromSegments implementations for the Rust standard library std::path::PathBuf type. PathBuf is a type for representing a file path in the operating system. This implementation is very useful for serving a static file from the Rocket application.

You might think serving from a specific path is dangerous because any person can try a path traversal such as "../../../password.txt". Fortunately, the FromSegments implementation for PathBuf has already thought about the security problem. As a result, access to sensitive paths has been disabled, for example, "..", ".", or "*".

Another segment type is the **ignored segment**. The ignored segment is defined as <_> or <_..>. If you declare an ignored segment, it will not show in the function argument list. You must declare ignored multiple segments as the last argument in a path, just like regular multiple segments.

An ignored segment is useful if you want to build an HTTP path that matches a lot of things, but you don't want to process it. For example, if you have the following lines of code, you can have a website that handles any path. It will handle /, /some, /1/2/3, or anything else:

```
#[get("/<_>")]
fn index() {}


#[launch]
fn rocket() -> Rocket<Build> {
    rocket::build().mount("/", routes![index])
}
```

Query

Just like a path, a query segment can be a static segment or dynamic segment (such as `"?<query1>&<query2>"`) or can be in a multiple `"?<query..>"` form. The multiple query form is called **trailing parameters**. Unlike the path, the query part does not have the ignored part, `"?<_>"`, or ignored trailing parameters such as `"?<_..>"`.

Neither dynamic queries nor trailing parameters are supposed to implement `FromParam`, but both must implement `rocket::form::FromForm` instead. We will discuss implementing `FromForm` more in *Chapter 8, Serving Static Assets and Templates*.

Rank

The path and query segments in a URI can be grouped into three **colors**: **static**, **partial**, or **wild**. If all segments of the path are static, the path is called a static path. If all segments of the query are static, we say the query has a static color. If all segments of the path or query are dynamic, we call the path or query wild. The partial color is when a path or query has both static and dynamic segments.

Why do we need these colors? They are required to determine the next parameter of the route, which is the **rank**. If we have multiple routes handling the same path, then Rocket will rank the functions and *start checking from the rank with the lowest number*. Let's see an example:

```
#[get("/<rank>", rank = 1)]
fn first(rank: u8) -> String {
    let result = rank + 10;
    format!("Your rank is, {}!", result)
}


#[get("/<name>", rank = 2)]
fn second(name: &str) -> String {
    format!("Hello, {}!", name)
}


#[launch]
fn rocket() -> Rocket<Build> {
    rocket::build().mount("/", routes![first, second])
}
```

Here, we see we have two functions handling the same path, but with two different function signatures. Since Rust does not support function overloading, we created the functions with two different names. Let's try calling each of the routes:

```
> curl http://127.0.0.1:8000/1
Your rank is, 11!


> curl http://127.0.0.1:8000/jane
Hello, jane!
```

When we look at the application log in the other terminal, we can see how Rocket chose the route:

```
GET /1:
   >> Matched: (first) GET /<rank>
   >> Outcome: Success
   >> Response succeeded.
GET /jane:
   >> Matched: (first) GET /<rank>
   >> `rank: u8` param guard parsed forwarding with error
      "jane"
   >> Outcome: Forward
   >> Matched: (second) GET /<name> [2]
   >> Outcome: Success
   >> Response succeeded.
```

Try reversing the rank in the source code and think about what would happen if you called it with u8 as the parameter. After that, try requesting the endpoint to see whether your guess is correct.

Let's recall Rocket's URI colors. Rocket ranks the colors of both path and query as follows:

- static path, static query = -12
- static path, partial query = -11
- static path, wild query = -10
- static path, none query = -9
- partial path, static query = -8
- partial path, partial query = -7

- partial path, wild query = -6
- partial path, none query = -5
- wild path, static query = -4
- wild path, partial query = -3
- wild path, wild query = -2
- wild path, none query = -1

You can see the path has a lower rank, static is lower than partial, and finally, partial is lower than the wild color. Keep this in mind when you create multiple routes, as the output might not be what you expect because your route may have a lower or higher ranking.

Format

Another parameter we can use in a route is `format`. In requests with the HTTP method with a payload, such as POST, PUT, PATCH and DELETE, the HTTP request Content-Type is checked against the value of this parameter. When handling HTTP requests without payloads, such as GET, HEAD, and OPTIONS, Rocket checks and matches the route's format with the HTTP requests' Accept header.

Let's create an example for the `format` parameter. Create a new application and add the following lines:

```
#[get("/get", format = "text/plain")]
fn get() -> &'static str {
    "GET Request"
}


#[post("/post", format = "form")]
fn post() -> &'static str {
    "POST Request"
}


#[launch]
fn rocket() -> Rocket<Build> {
    rocket::build().mount("/", routes![get, post])
}
```

If you pay attention closely, the format for the `/get` endpoint uses the `"text/plain"` IANA (Internet Assigned Numbers Authority) media type, but the format for the `/post` endpoint is not the correct IANA media type. This is because Rocket accepts the following shorthand and converts them to the correct IANA media type:

- `"any"` → `"*/*"`
- `"binary"` → `"application/octet-stream"`
- `"bytes"` → `"application/octet-stream"`
- `"html"` → `"text/html; charset=utf-8"`
- `"plain"` → `"text/html; charset=utf-8"`
- `"text"` → `"text/html; charset=utf-8"`
- `"json"` → `"application/json"`
- `"msgpack"` → `"application/msgpack"`
- `"form"` → `"application/x-www-form-urlencoded"`
- `"js"` → `"application/javascript"`
- `"css"` → `"text/css; charset=utf-8"`
- `"multipart"` → `"multipart/form-data"`
- `"xml"` → `"text/xml; charset=utf-8"`
- `"pdf"` → `"application/pdf"`

Now, run the application and call each of the two endpoints to see how they behave. First, call the `/get` endpoint with both the correct and incorrect `Accept` header:

```
> curl -H "Accept: text/plain" http://127.0.0.1:8000/get
GET Request

> curl -H "Accept: application/json" http://127.0.0.1:8000/get

{
    "error": {
      "code": 404,
      "reason": "Not Found",
      "description": "The requested resource could not be
      found."
```

```
    }
}
```

The request with the correct `Accept` header returns the correct response, while the request with the incorrect `Accept` header returns `404` but with a `"Content-Type: application/json"` response header. Now, send the `POST` requests to the `/post` endpoint to see the responses:

```
> curl -X POST -H "Content-Type: application/x-www-form-
urlencoded" http://127.0.0.1:8000/post

POST Request

> curl -X POST -H "Content-Type: text/plain"
http://127.0.0.1:8000/post

<!DOCTYPE html>
<html lang="en">
<head>
    <meta charset="utf-8">
    <title>404 Not Found</title>
</head>
...
</html>
```

Our application outputs the expected response, but the `Content-Type` of the response is not what we expected. We will learn how to create a default error handler later in this chapter.

Data

The `data` parameter in the route is for processing the request body. The data must be in a dynamic form such as a dynamic `<something>` URI segment. After that, the declared attribute must be included as a parameter in the function following the route attribute. For example, look at the following lines:

```
#[derive(FromForm)]
struct Filters {
    age: u8,
```

```
        active: bool,
    }


    #[post("/post", data = "<data>")]
    fn post(data: Form<Filters>) -> &'static str {
        "POST Request"
    }


    #[launch]
    fn rocket() -> Rocket<Build> {
        rocket::build().mount("/", routes![post])
    }
```

If you do not include data as a parameter in the function following the route, Rust will complain about it at compile time. Try removing the data parameters in the function signature and try compiling it to see the compiler error output in action.

We will learn more about data later when we implement forms and upload files to the server. Now that we have learned about Rocket routes, let's make an application to implement a route that handles a request.

Implementing route handlers

Here, we will make an application that handles a route. We are reusing the first code that we wrote in this chapter. The idea is that we have several user data, and we want to send requests that will select and return the selected user data according to the ID sent in the request. In this part, we will implement the request and selecting part of the route handlers. In the next section, we will learn how to create a custom response type. In the subsequent section, we will create a handler for when the request does not match any user data we have. And finally, in the last section, we will create a default error handler to handle invalid requests.

Let's start by copying the first code into a new folder. After that, in src/main.rs, add a User struct after the Filter definition:

```
    struct Filters {
        ...
    }


    #[derive(Debug)]
```

```
struct User {
    uuid: String,
    name: String,
    age: u8,
    grade: u8,
    active: bool,
}
```

For the `User` struct, we are using `uuid` for object identification. The reason is that if we use `usize` or another numeric type as an ID without any authentication, we might fall into the **Insecure Direct Object References** (**IDOR**) security vulnerability where an unauthorized user can easily guess any number as an ID. The UUID as an identifier is harder to guess.

Also, in real-world applications, we should probably create transparent encryption for the *name* and *age* as such information can be considered personally identifiable information, but let's skip it in this book for the sake of learning.

We also add the `#[derive(Debug)]` attribute on top of the struct. The attribute automatically creates an implementation for the struct to be printed using `fmt::Debug`. We can then use it in code, such as `format!("{:?}", User)`. One of the requirements for the `Debug` attribute is that all type members must implement `Debug` as well; however, this is not a problem in our case, as all Rust standard library types already implement the `Debug` trait.

As for the next step, we want to store several `User` data in a collection data structure. We can store them in a `[User; 5]` array or a growable `std::vec::Vec` array type. To find user data inside the array, we can iterate the array or Vec one by one until the end or until a match is found, but this is not ideal as it is time-consuming for a large array.

In computer science, there are better data structures in which we can store data and easily find objects by their index, such as a hash map. Rust has many libraries that implement various data structures, and a hash map is one of them. In the standard library, we can find it in `std::collections::HashMap`.

Besides using a standard library, we can use other alternatives, since the Rust community has already created a lot of data structure-related libraries. Try searching in `https://crates.io` or `https://lib.rs`. For example, if we are not using the standard library, we can use an alternative crate such as `hashbrown`.

Let's implement it in our `src/main.rs` file after the `User` struct declaration. Unfortunately, the `HashMap` creation requires heap allocation, so we cannot assign `HashMap` to a static variable. Adding the following code will not work:

```
use std::collections::HashMap;

...

static USERS: HashMap<&str, User> = {
    let map = HashMap::new();
    map.insert(
        "3e3dd4ae-3c37-40c6-aa64-7061f284ce28",
        User {
            uuid: String::from("3e3dd4ae-3c37-40c6-aa64-
            7061f284ce28"),
            name: String::from("John Doe"),
            age: 18,
            grade: 1,
            active: true,
        },
    );
    map
};
```

There are several ways to assign `HashMap` to a static variable, but the best suggestion is to use the `lazy_static!` macro from the `lazy_static` crate, which runs the code at runtime and performs heap allocation. Let's add it to our code. First, add `lazy_static` in the `Cargo.toml` dependencies:

```
[dependencies]
lazy_static = "1.4.0"
rocket = "0.5.0-rc.1"
```

After that, use and implement it in the code as follows. Feel free to add extra users if you want to test it later:

```
use lazy_static::lazy_static;
use std::collections::HashMap;

...

lazy_static! {
    static ref USERS: HashMap<&'static str, User> = {
```

```
        let mut map = HashMap::new();
        map.insert(
            "3e3dd4ae-3c37-40c6-aa64-7061f284ce28",
            User {
                ...
            },
        );
        map
    };
}
```

Let's modify fn user(...) as follows:

```
#[get("/user/<uuid>", rank = 1, format = "text/plain")]
fn user(uuid: &str) -> String {
    let user = USERS.get(uuid);
    match user {
        Some(u) => format!("Found user: {:?}", u),
        None => String::from("User not found"),
    }
}
```

We want the function to return something when we call it, therefore, we add -> String in the function signature.

HashMap has many methods, such as insert() for inserting a new key and value, or keys(), which returns an iterator for the keys in HashMap. We are just using get(), which returns std::option::Option. Remember, Option is just an enum, which can be None, or Some(T) if it contains a value. Finally, the match control flow operator returns a string appropriately depending on whether the value is None or Some(u).

Now, if we try to send a GET request to http://127.0.0.1:8000/ user/3e3dd4ae-3c37-40c6-aa64-7061f284ce28, we can see that it will return the correct response, and if we send a GET request to http://127.0.0.1:8000/ user/other, it will return "User not found".

Now, let's implement the users() function. Let's recall the original signature:

```
#[route(GET, uri = "/users/<grade>?<filters..>")]
fn users(grade: u8, filters: Filters) {}
```

Because u8 already implements FromParam, we can just use it as it is. But, we want to see how we can implement FromParam for a custom type. Let's change our use case to have a path such as "/users/<name_grade>?<filters...>".

First, create a custom NameGrade struct. The 'r annotation means that this struct should only live as long as the referenced string in its name field:

```
struct NameGrade<'r> {
    name: &'r str,
    grade: u8,
}
```

If we want to implement a trait, we have to look at the signature of that trait. The Rust compiler requires a type to implement all methods and a type placeholder in a trait. We can find the trait definition for FromParam from the Rocket API documentation:

```
pub trait FromParam<'a>: Sized {
    type Error: Debug;
    fn from_param(param: &'a str) -> Result<Self, Self::Error>;
}
```

The type Error: Debug; is called a type placeholder. Some traits require the implementation to have a certain type. Any type that implements this trait should use a concrete type, which also has a debug trait. Because we just want to show an error message, we can use &'static str as the Error type for this implementation. Then, write the trait implementation signature for NameGrade as follows:

```
use rocket::{request::FromParam, Build, Rocket};
...
impl<'r> FromParam<'r> for NameGrade<'r> {
    type Error = &'static str;
    fn from_param(param: &'r str) -> Result<Self, Self::
    Error> {}
}
```

Inside the function, add the message that we want to show to the app user:

```
const ERROR_MESSAGE: Result<NameGrade, &'static str> =
Err("Error parsing user parameter");
```

Then, let's split the input parameter at the '_' character:

```
let name_grade_vec: Vec<&'r str> = param.split('_').collect();
```

The name_grade_vec length will be either 2 or other, so we can use match on it. As name_grade_vec[0] is a string, we can use it as it is, but for the second member, we have to parse it. And, since the result can be anything, we have to use a special syntax that is formed as in ::<Type>. This syntax is fondly called **turbofish** by the Rust community.

Just like Option, Result is just an enum which can either be Ok(T) or Err(E). If the program successfully parses u8, the method can return Ok(NameGrade{...}), or else the function can return Err("..."):

```
match name_grade_vec.len() {
    2 => match name_grade_vec[1].parse::<u8>() {
        Ok(n) => Ok(Self {
            name: name_grade_vec[0],
            grade: n,
        }),
        Err(_) => ERROR_MESSAGE,
    },
    _ => ERROR_MESSAGE,
}
```

Now that we have implemented FromParam for NameGrade, we can use NameGrade as the parameter in the users() function. We also want String as the return type of the function:

```
#[get("/users/<name_grade>?<filters..>")]
fn users(name_grade: NameGrade, filters: Filters) -> String {}
```

Inside the function, write the routine that filters the USERS hash map with name_grade and filters:

```
let users: Vec<&User> = USERS
    .values()
    .filter(|user| user.name.contains(&name_grade.name) &&
    user.grade == name_grade.grade)
    .filter(|user| user.age == filters.age && user.active
    == filters.active)
    .collect();
```

HashMap has the `values()` method, which returns `std::collections::hash_map::Values`. `Values` implements `std::iter::Iterator`, so we can filter it using the `filter()` method. The `filter()` method accepts a *closure*, which returns Rust bool type. The `filter()` method itself returns `std::iter::Filter`, which implements the `Iterator` trait. The `Iterator` trait has the `collect()` method, which can be used to collect items into the collection. Sometimes, you have to use the `::<Type>` turbofish in `collect::<Type>()` if the result type cannot be inferred by the compiler.

After that, we can convert the collected users into `String`:

```
if users.len() > 0 {
    users
        .iter()
        .map(|u| u.name.to_string())
        .collect::<Vec<String>>()
        .join(",")
} else {
    String::from("No user found")
}
```

After this is done, run the application and try calling the `users()` function:

```
curl -G -d age=18 -d active=true http://127.0.0.1:8000/users/
John_1
```

It works, but the problem with this is that the query parameters are cumbersome; we want `Filters` to be optional. Let's modify the code a little bit. Change the signature of `fn users` to the following:

```
fn users(name_grade: NameGrade, filters: Option<Filters>) ->
String {
    ...
            .filter(|user| {
                if let Some(fts) = &filters {
                    user.age == fts.age && user.active ==
                    fts.active
                } else {
                    true
                }
```

```
        })
    ...
```

You might be confused by this piece of code: `if let Some(fts) = &filters`. It's one of the destructuring syntaxes in Rust, just like this piece of code:

```
match something {
    Ok(i) => /* use i here */ "",
    Err(err) => /* use err here */ "",
}
```

We have implemented the requests part for these two endpoints, `user()` and `users()`, but the return type for those two endpoints is the Rust standard library type, `String`. We want to use our own custom type. So, let's see how we can create a response from the `User` struct directly in the next section.

Creating responses

Let's implement a custom response for the `User` type. In Rocket, all types that implement `rocket::response::Responder` can be used as a return type in a function that handles routes.

Let's take a look at the signature of the `Responder` trait. This trait requires two lifetimes, `'r` and `'o`. The result `'o` lifetime must at least be equal to the `'r` lifetime:

```
pub trait Responder<'r, 'o: 'r> {
    fn respond_to(self, request: &'r Request<'_>) ->
    Result<'o>;
}
```

First, we can include the required module to be used for implementing the `Responder` trait for the `User` struct:

```
use rocket::http::ContentType;
use rocket::response::{self, Responder, Response};
use std::io::Cursor;
```

After that, add the implementation signature for the `User` struct:

```
impl<'r> Responder<'r, 'r> for &'r User {
    fn respond_to(self, _: &'r Request<'_>) ->
```

```
    response::Result<'r> {        }
}
```

Why do we use `rocket::response::{self...}` instead of `rocket::response::{Result...}`? If we return `-> Result`, we cannot use `std::result::Result`, which is a pretty ubiquitous type in Rust. Write the following lines in the `respond_to()` method body:

```
let user = format!("Found user: {:?}", self);
Response::build()
    .sized_body(user.len(), Cursor::new(user))
    .raw_header("X-USER-ID", self.uuid.to_string())
    .header(ContentType::Plain)
    .ok()
```

The application generates a user `String` from the `User` object, then generates `rocket::response::Builder` by calling `Response::build()`. We can set various payloads for the `Builder` instance; for example, the `sized_body()` method adds the body of the response, `raw_header()` and `header()` add the HTTP header, and finally, we generate `response::Result()` using the `finalize()` method.

The first parameter of the `sized_body()` method is `Option`, and the parameter can be `None`. As a result, the `sized_body()` method requires the second parameter to implement the `tokio::io::AsyncRead + tokio::io::AsyncSeek` trait to automatically determine the size. Fortunately, we can wrap the body in `std::io::Cursor` as Tokio already implements those traits for `Cursor`.

There's a common pattern that we can observe when we implement the `std::iter::Iterator` trait and `rocket::response::Builder`, called the **builder pattern**. It's a pattern used to generate an instance by chaining methods. Take a look at the following example. We can then generate a `Something` instance by chaining the command, such as `Something.new().func1().func2()`:

```
struct Something {}

impl Something {
    fn new() -> Something { ... }
    fn func1(&mut self) -> &mut Something { ... }
    fn func2(&mut self) -> &mut Something { ... }
}
```

Let's also modify the users() function to return a new Responder. We are defining a new type, which is commonly known as a **newtype** idiom. This idiom is useful if we want to wrap a collection or work around **orphan rules**.

Orphan rules mean neither type nor impl are in our application or crate. For example, we cannot implement impl Responder for Iterator in our application. The reason is that Iterator is defined in the standard library, and Responder is defined in the Rocket crate.

We can use the newtype idiom as in the following line:

```
struct NewUser<'a>(Vec<&'a User>);
```

Notice that the struct has a **nameless field**. The form for a struct with a nameless field is as follows: struct NewType(type1, type2, ...).

We can also call a struct with a nameless field a **tuple struct** because the field is like a **tuple** (type1, type2, type3). We then can access the struct's field by its index, such as self.0, self.1, and so on.

After the newtype definition, add the implementation as follows:

```
impl<'r> Responder<'r, 'r> for NewUser<'r> {
    fn respond_to(self, _: &'r Request<'_>) ->
    response::Result<'r> {
        let user = self
            .0
            .iter()
            .map(|u| format!("{:?}", u))
            .collect::<Vec<String>>()
            .join(",");
        Response::build()
            .sized_body(user.len(), Cursor::new(user))
            .header(ContentType::Plain)
            .ok()
    }
}
```

Like the `Responder` implementation for the `User` type, in the `Responder` implementation for `NewUser`, we basically iterate a collection of users again, collect them as a String, and build the `response::Result` again.

Finally, let's use both the `User` and `NewUser` structs as a response type in the `user()` and `users()` functions:

```
#[get("/user/<uuid>", rank = 1, format = "text/plain")]
fn user(uuid: &str) -> Option<&User> {
    let user = USERS.get(uuid);
    match user {
        Some(u) => Some(u),
        None => None,
    }
}


#[get("/users/<name_grade>?<filters..>")]
fn users(name_grade: NameGrade, filters: Option<Filters>) ->
Option<NewUser> {
    ...
    if users.len() > 0 {
        Some(NewUser(users))
    } else {
        None
    }
}
```

Now that we've learned how to implement the `Responder` trait for a type, let's learn more about the wrappers that Rocket offers in the next section.

Wrapping Responder

Rocket has two modules that can be used to wrap the returned `Responder`.

The first module is `rocket::response::status`, which has the following structs: `Accepted`, `BadRequest`, `Conflict`, `Created`, `Custom`, `Forbidden`, `NoContent`, `NotFound`, and `Unauthorized`. All the responders except `Custom` set the status just like their corresponding HTTP response code. For example, we can modify the previous `user()` function as follows:

```
use rocket::response::status;

...

fn user(uuid: &str) -> status::Accepted<&User>  {

    ...

    status::Accepted(user)

}
```

The `Custom` type can be used to wrap a response with other HTTP code not available in the other structs. For example, take a look at the following lines:

```
use rocket::http::Status;
use rocket::response::status;

...

fn user(uuid: &str) -> status::Custom<&User>  {

    ...

    status::Custom(Status::PreconditionFailed, user)

}
```

The other module, `rocket::response::content`, has the following structs: `Css`, `Custom`, `Html`, `JavaScript`, `Json`, `MsgPack`, `Plain`, and `Xml`. Like the `status` module, the `content` module is used to set `Content-Type` of the response. For example, we can modify our code to the following lines:

```
use rocket::response::content;
use rocket::http::ContentType;

...

fn user(uuid: &str) -> content::Plain<&User> {

    ...

    content::Plain(user)

}
```

```
. . .
fn users(name_grade: NameGrade, filters: Option<Filters>) ->
content::Custom<NewUser> {
    . . .
    status::Custom(ContentType::Plain, NewUser(users));
}
```

We can also combine both modules as in the following example:

```
fn user(uuid: &str) -> status::Accepted<content::Plain<&User>>
{
    . . .
    status::Accepted(content::Plain(user))
}
```

We can rewrite this using `rocket::http::Status` and `rocket::http::ContentType`:

```
fn user(uuid: &str) -> (Status, (ContentType, &User)) {
    . . .
    (Status::Accepted, (ContentType::Plain, user))
}
```

Now, you might be wondering how those structs can create HTTP `Status` and `Content-Type` and use another `Responder` implementor body. The answer is because the `Response` struct has two methods: `join()` and `merge()`.

Let's assume there are two `Response` instances: `original` and `override`. The `original.join(override)` method merges the `override` body and status if it's not already present in `original`. The `join()` method also adjoins the same headers from `override`.

Meanwhile, the `merge()` method replaces the `original` body and status with that of `override`, and replaces the `original` header if it exists in `override`.

Let's rewrite our application to use a default response. This time we want to add a new HTTP header, `"X-CUSTOM-ID"`. To do so, implement the following function:

```
fn default_response<'r>() -> response::Response<'r> {
    Response::build()
        .header(ContentType::Plain)
        .raw_header("X-CUSTOM-ID", "CUSTOM")
```

```
        .finalize()
}
```

Then, modify the `Responder` implementation for the `User` struct:

```
fn respond_to(self, _: &'r Request<'_>) -> response::Result<'r>
{
    let base_response = default_response();
    let user = format!("Found user: {:?}", self);
    Response::build()
        .sized_body(user.len(), Cursor::new(user))
        .raw_header("X-USER-ID", self.uuid.to_string())
        .merge(base_response)
        .ok()
}
```

Finally, modify the `Responder` implementation for `NewUser`. But this time, we want to add extra values: the `"X-CUSTOM-ID"` header. We can do that using the `join()` method:

```
fn respond_to(self, _: &'r Request<'_>) -> response::Result<'r>
{
    let base_response = default_response();
    ...
    Response::build()
        .sized_body(user.len(), Cursor::new(user))
        .raw_header("X-CUSTOM-ID", "USERS")
        .join(base_response)
        .ok()
}
```

Try opening the URL for `user` and `users` again; you should see the correct `Content-Type` and `X-CUSTOM-ID`:

```
< x-custom-id: CUSTOM
< content-type: text/plain; charset=utf-8
< x-custom-id: USERS
< x-custom-id: CUSTOM
< content-type: text/plain; charset=utf-8
```

Built-in implementations

Besides `content` and `status` wrappers, Rocket already implemented the `Responder` trait for several types to make it easier for developers. The following is a list of types already implementing the `Responder` trait:

- `std::option::Option` – We can return `Option<T>` for any type of `T` that already implements `Responder`. If the returned variant is `Some(T)`, then `T` is returned to the client. We already see the example of this return type in the `user()` and `users()` functions.

- `std::result::Result` – Both variants `T` and `E` in `Result<T, E>` should implement `Responder`. For example, we can change our `user()` implementation to return `status::NotFound` as in the following lines:

```
use rocket::response::status::NotFound;

...

fn user(uuid: &str) -> Result<&User, NotFound<&str>> {
    let user = USERS.get(uuid);
    user.ok_or(NotFound("User not found"))
}
```

- `&str` and `String` – These types are returned with the text content as the response body and `Content-Type` `"text/plain"`.

- `rocket::fs::NamedFile` – This `Responder` trait automatically returns a file specified with `Content-Type` based on the file content. For example, we have the `"static/favicon.png"` file and we want to serve it in our application. Take a look at the following example:

```
use rocket::fs::{NamedFile, relative};
use std::path::Path;

#[get("/favicon.png")]
async fn favicon() -> NamedFile {
    NamedFile::open(Path::new(relative!(
    "static")).join("favicon.png")).await.unwrap()
}
```

- `rocket::response::Redirect` – `Redirect` is used to return a `redirect` response to the client. We will discuss `Redirect` more in *Chapter 8*, *Serving Static Assets and Templates*.

- `rocket_dyn_templates::Template` – This responder returns a dynamic template. We will discuss templating more in *Chapter 8*, *Serving Static Assets and Templates*.

- `rocket::serde::json::Json` – This type makes it easy to return the JSON type. To use this responder implementation, you must enable the `"json"` feature in `Cargo.toml` as follows: `rocket = {version = "0.5.0-rc.1",` `features = ["json"]}`. We will discuss more about JSON in *Chapter 11*, *Securing and Adding an API and JSON*.

- `rocket::response::Flash` – `Flash` is a type of cookie that will be erased after a client accesses it. We will learn how to use this type in *Chapter 11*, *Securing and Adding an API and JSON*.

- `rocket::serde::msgpack::MsgPack` – **MessagePack** is a binary serialization format just like JSON. To use this, you must enable the `"msgpack"` feature in `Cargo.toml`.

- Various `stream` responders in the `rocket::response::stream` module – We will learn more about these responders in *Chapter 9*, *Displaying Users' Post*, and *Chapter 10*, *Uploading and Processing Posts*.

We have already implemented a couple of routes, derived `FromParam`, and created types that implemented the `Responder` trait. In the next section, we will learn how to make a default error catcher for the same kind of HTTP status code.

Making default error catchers

An application should be able to handle an error that may occur anytime during processing. In a web application, the standardized way to return an error to a client is by using an HTTP status code. Rocket provides a way to handle returning errors to clients in the form of `rocket::Catcher`.

The catcher handler works just like a route handler, with a few exceptions. Let's modify our last application to see how it works. Let's recall how we implemented the `user()` function:

```
fn user(uuid: &str) -> Result<&User, NotFound<&str>> {
    let user = USERS.get(uuid);
    user.ok_or(NotFound("User not found"))
}
```

If we request GET /user/wrongid, the application will return an HTTP response with code 404, a "text/plain" content type, and a "User not found" body. Let's change the function back to the return Option:

```
fn user(uuid: &str) -> Option<&User> {
    USERS.get(uuid)
}
```

A function returning Option where the variant is None will use the default 404 error handler. After that, we can implement the default 404 handler as follows:

```
#[catch(404)]
fn not_found(req: &Request) -> String {
    format!("We cannot find this page {}.", req.uri())
}
```

Notice the #[catch(404)] attribute above the function. It looks like a route directive. We can use any valid HTTP status code between 200 to 599 or default. If we put default, it will be used for any HTTP status code not declared in the code.

Like route, the catch attribute must be put above a free function. We cannot put the catch attribute above a method inside the impl block. Also like the route handling function, the catcher function must return a type that implements Responder.

The function handling an error can have zero, one, or two parameters. If the function has one parameter, the parameter type must be &rocket::Request. If the function has two parameters, the first parameter type must be rocket::http::Status and the second parameter must be &Request.

The way the catcher function connects to Rocket is a little bit different. Where we use mount() and the routes! macro for the route handling functions, we use register() and the catchers! macro for catcher functions:

```
#[launch]
fn rocket() -> Rocket<Build> {
    rocket::build().mount("/", routes![user, users,
    favicon]).register("/", catchers![not_found])
}
```

How can we tell a route handling function to use a catcher? Let's say a catcher has been defined and registered as in the following lines:

```
#[catch(403)]
```

```
fn forbidden(req: &Request) -> String {
    format!("Access forbidden {}.", req.uri())
}


fn rocket() -> Rocket<Build> {
    rocket::build().mount("/", routes![user, users,
    favicon]).register("/", catchers![not_found,
    forbidden])
}
```

We can then return rocket::http::Status directly on the route handling function. The status will then be forwarded to any registered catcher or Rocket built-in catcher:

```
use rocket::http::{Status, ContentType};
...
fn users(name_grade: NameGrade, filters: Option<Filters>) ->
Result<NewUser, Status> {
    ...
    if users.is_empty() {
        Err(Status::Forbidden)
    } else {
        Ok(NewUser(users))
    }
}
```

Try calling the GET request to this endpoint and see what happens:

```
curl -v http://127.0.0.1:8000/users/John_2
...
< HTTP/1.1 403 Forbidden
< content-type: text/plain; charset=utf-8
...
<
* Connection #0 to host 127.0.0.1 left intact
Access Forbidden /users/John_2.* Closing connection 0
```

The application returns the string from the 403 default handler and the correct HTTP status is also returned.

Summary

This chapter explored one of the most important parts of the Rocket framework. We learned about a route and its parts such as HTTP methods, URIs, path, query, rank, and data. We also implemented a couple of routes and various types related to routes in an application. After that, we explored ways to create responder types and learned about various wrappers and types already implemented in the `Responder` trait. Finally, we learned how to create a catcher and connect it to a Rocket application.

In the next chapter, we will learn about other Rocket components such as states and fairings. We will learn the initialization process of a Rocket application, and how we can use those states and fairings to create more modern and complex applications.

4

Building, Igniting, and Launching Rocket

Many web applications require some kind of object management that can be reused again and again, be it a connection pool for a database server, a connection to a memory store, an HTTP client to third-party servers, or any other object. Another common feature in a web application is **middleware**.

In this chapter, we will discuss two Rocket features (state and fairings), which act as the reusable object management and middleware parts of Rocket. We will also learn how to create and use connections to database servers, which is very important in almost all web applications.

After completing this chapter, we expect you to be able to use and implement the reusable object management and middleware parts of the Rocket web framework. We also expect you to be able to connect to a database of your own choice.

In this chapter, we're going to cover the following main topics:

- Managing state
- Working with a database
- Attaching Rocket fairings

Technical requirements

Besides the usual requirements of a Rust compiler, a text editor, and an HTTP client, starting from this chapter, we're going to work with a database. The database we're going to use throughout this book is PostgreSQL, and you can download it from `https://www.postgresql.org/`, install it from your operating system package manager, or use a third-party server such as **Amazon Web Services (AWS)**, Microsoft Azure, or **Google Cloud Platform (GCP)**.

We're going to see how to connect to other **Relational Database Management Systems (RDBMSs)** such as SQLite, MySQL, or Microsoft SQL Server, and you can adjust the lesson code to make the type suitable to those RDBMSs, but it's easier to follow using PostgreSQL.

You can find the source code for this chapter at `https://github.com/PacktPublishing/Rust-Web-Development-with-Rocket/tree/main/Chapter04`.

Managing state

In a web application, usually, programmers need to create an object that can be reused during the request/response life cycle. In the Rocket web framework, that object is called a **state**. A state can be anything such as a database connection pool, an object to store various customer statistics, an object to store a connection to a memory store, a client to send **Simple Mail Transfer Protocol (SMTP)** emails, and many more.

We can tell Rocket to maintain the state, and this is called a **managed state**. The process of creating a managed state is quite simple. We need to initialize an object, tell Rocket to manage it, and finally use it in a route. One caveat is that we can manage many states from different types, but Rocket can only manage one instance of a Rust type.

Let's try it directly. We are going to have a visitor counter state and tell Rocket to manage it and increment the counter for every incoming request. We can reuse the previous application from the previous chapter, copy the program from `Chapter03/15ErrorCatcher` into `Chapter04/01State`, and rename the application in `Cargo.toml` as `chapter4`.

In `src/main.rs`, define a struct to hold the value of the visitor counter. For the state to work, the requirement is `T: Send + Sync + 'static`:

```
use std::sync::atomic::AtomicU64;

...

struct VisitorCounter {
    visitor: AtomicU64,
}
```

We already know that `'static` is a lifetime marker, but what is `Send + Sync`?

In modern computing, due to its complexity, there are many ways a program can be executed in a way not intended. For example, multithreading makes it hard to know whether a variable value has been changed on another thread or not. Modern CPUs also perform branch prediction and execute multiple instructions at the same time. Sophisticated compilers also rearrange the resulting binary code execution flow to optimize the result. To overcome those problems, some kind of syncing is needed in the Rust language.

The Rust language has traits and memory containers to solve syncing problems depending on how programmers intended the application to work. We might want to create an object in the heap and share the reference to that object in multiple other objects. For example, we create object x, and we use the reference of x, `&x` in other objects field, `y` and `z`. This creates another problem, as the program can delete x in other routines making the program unstable. The solution is to create different containers for different use cases. These include `std::cell::Rc` and `std::box::Box`, among others.

`std::marker::Send` is one of those traits. The `Send` trait is making sure any T type is safe to be transferred to another thread. Almost all types in the `std` library are `Send`, with a few exceptions such as `std::rc::Rc` and `std::cell::UnsafeCell`. Rc is a single-threaded reference counted pointer.

Meanwhile, `std::marker::Sync` is saying the T type is safe to be shared across multiple threads. That only holds `true` if the `&T` reference is safe to be sent to another thread. Not all `Send` types are `Sync`. For example, `std::cell::Cell` and `std::cell::RefCell` are `Send` but not `Sync`.

Both `Send + Sync` are **marker traits**; they don't have a function to be implemented by a type. They are also **unsafe traits,** so all types implementing these manually are also unsafe and can lead to undefined behavior. So, how do we implement `Send + Sync` in our type? These types are also **automatically-derived traits**, which means that a type in which all members are implementing `Send` automatically becomes a `Send` type. Almost all types in the `std` library are `Send + Sync` apart from the raw pointers, `Rc`, `Cell`, and `RefCell`.

What is AtomicU64? With the regular u64 type, even though it's Send + Sync, there's no synchronization between threads so a data race condition might happen. For example, two threads access the same variable, x (which has a value of 64) at the same time, and they increment the value by one. We expect the result to be 66 (as there are two threads), but because there's no synchronization between threads, the final result is unpredictable. It can be 65 or 66.

The types in the std::sync module provide a couple of ways to share updates between multiple threads, including the std::sync::Mutex, std::sync::RwLock, and std::sync::atomic types. We can also use other libraries that might provide better speed than a standard library, such as the parking_lot crate.

Now we have defined VisitorCounter, let's initialize it and tell Rocket to manage it as a state. Write the code inside the rocket() function as in the following lines:

```
fn rocket() -> Rocket<Build> {
    let visitor_counter = VisitorCounter {
        visitor: AtomicU64::new(0),
    };
    rocket::build()
        .manage(visitor_counter)
        .mount("/", routes![user, users, favicon])
        .register("/", catchers![not_found, forbidden])
}
```

After we tell Rocket to manage the state, we can use it inside the route handling functions. In the previous chapter, we learned about dynamic segments that we have to use in function arguments. There are other arguments we can use in a route handling function, which we call **request guards**. They are called *guards* because if a request does not pass the validation inside the guard, the request will be rejected, and an error response will be returned.

Any type that implements rocket::request::FromRequest can be considered a request guard. Incoming requests are then validated against each request guard from left to right and will short circuit and return an error if the request is not valid.

Suppose we have a route handling function as in the following lines:

```
#[get("/<param1>")]
fn handler1(param1: u8, type1: Guard1, type2: Guard2) {}
```

The Guard1 and Guard2 types are the request guards. The incoming request is then validated against the Guard1 methods, and if an error occurs, the proper error response will be returned immediately.

We will learn about and implement request guards throughout the book, but we will just use a request guard without implementing it in this chapter. FromRequest is already implemented for rocket::State<T>, so we can use it in the route handling function.

Now that we have learned why we use State in a route handling function, let's use it in our functions. We want to set the visitor counter so each hit to the request should increment the counter:

```
use rocket::{Build, Rocket, State};


#[get("/user/<uuid>", rank = 1, format = "text/plain")]
fn user<'a>(counter: &State<VisitorCounter>, uuid: &'a str) ->
Option<&'a User> {
    counter.visitor.fetch_add(1, Ordering::Relaxed);
    println!("The number of visitor is: {}", counter.
    visitor.load(Ordering::Relaxed));
    USERS.get(uuid)
}
```

Why do we add the 'a lifetime? We are adding a new reference argument, and Rust cannot infer which lifetime the returned &User should follow. In this case, we are saying the lifetime of the User reference should be as long as uuid.

Inside the function, we use the AtomicU64 fetch_add() method to increment the value of the visitor, and we print the value using the AtomicU64 load() method.

Let's add the same for the users() function, but since we have the exact same routine with the user() function, let's make another function instead:

```
impl VisitorCounter {
    fn increment_counter(&self) {
        self.visitor.fetch_add(1, Ordering::Relaxed);
        println!(
            "The number of visitor is: {}",
            self.visitor.load(Ordering::Relaxed)
        );
```

```
    }
}
...
fn user<'a>(counter: &State<VisitorCounter>, uuid: &'a str) ->
Option<&'a User> {
    counter.increment_counter();

    ...
}
...
fn users<'a>(
    counter: &State<VisitorCounter>,
    name_grade: NameGrade,
    filters: Option<Filters>,
) -> Result<NewUser<'a>, Status> {
    counter.increment_counter();

    ...
}
```

This example works fine with the `Atomic` type, but if you ever needed a more complex type to work with, such as `String`, `Vec`, or `Struct`, try using `Mutex` or `RwLock` from either the standard library or a third-party crate such as `parking_lot`.

Now that we know what `State` is in Rocket, let's expand our application by combining it with a database server. We will use `State` for storing the connection to the database.

Working with a database

Currently, in our application, we are storing user data in a static variable. This is very cumbersome, as it is inflexible and we cannot update the data easily. Most modern applications handling data will use some kind of persistent storage, be it filesystem-backed storage, a document-oriented database, or a traditional RDBMS.

Rust has many libraries to connect to various databases or database-like storage. There's the `postgres` crate, which works as a PostgreSQL client for Rust. There are also other clients such as `mongodb` and `redis`. For **object-relational mapping (ORM)** and Query Builder, there's `diesel`, which can be used to connect to various database systems. For connection pool management, there are the `deadpool` and `r2d2` crates. All crates have their strengths and limitations, such as not having an asynchronous application.

In this book, we're going to use `sqlx` to connect to an RDBMS. `sqlx` claims to be an SQL toolkit for Rust. It has abstractions for clients to connect to various RDBMSs, it has a connection pool trait, and it can also be used to convert types to queries and query responses to Rust types.

As mentioned in the *Technical requirements* section of this chapter, we're going to use PostgreSQL as our RDBMS, so please prepare the connection information to PostgreSQL.

After that, follow these steps to convert our application into using a database:

1. We will reuse our application again. The first thing we want to do is to install `sqlx-cli` by typing this command in the terminal:

    ```
    cargo install sqlx-cli
    ```

    `sqlx-cli` is a useful command-line application to create a database, create migrations, and run the migrations. It's not as sophisticated as migration tools in other established frameworks, but it does its job very well.

2. Prepare the connection information and set the `DATABASE_URL` environment variable in your terminal. The `DATABASE_URL` format should look as follows, depending on which RDBMS you are using:

    ```
    postgres://username:password@localhost:port/db_
    name?connect_options
    ```

    For `connect_options`, it's in query form, and the reference can be found at `https://www.postgresql.org/docs/current/libpq-connect.html#LIBPQ-CONNSTRING`. Other DATABASE_URL format for other RDBMS might look like:

    ```
    mysql://username:password@localhost:port/db_name
    ```

    Or `sqlite::memory:` or `sqlite://path/to/file.db?connect_options` or `sqlite:///path/to/file.db?connect_options`. The connect options for SQLite can be found at `https://www.sqlite.org/uri.html`.

3. Create a new database using this command:

    ```
    sqlx database create
    ```

4. We can create a migration named `create_users` using this command:

    ```
    sqlx migrate add create_users
    ```

5. The `sqlx` CLI will create a new folder named `migrations` inside the root directory of our application, and inside the folder, there will be a file with the `timestamp_migration_name.sql` pattern. In our example, the filename will look like `migrations/20210923055406_create_users.sql`. Inside the file, we can write SQL queries to create or modify the **data definition layer** (**DDL**). In this case, we want to make the content the same as the `User` struct, so let's write the following code into the SQL file:

```
CREATE TABLE IF NOT EXISTS users
(
    uuid    UUID PRIMARY KEY,
    name    VARCHAR NOT NULL,
    age     SMALLINT NOT NULL DEFAULT 0,
    grade   SMALLINT NOT NULL DEFAULT 0,
    active BOOL NOT NULL DEFAULT TRUE
);
CREATE INDEX name_active_idx ON users(name, active);
```

How do we know what the mapping between the database column type and Rust type is? `sqlx` provides its own mapping; we can find the documentation at `https://docs.rs/sqlx`. The crate has great modules for a supported database. We can search for it in the top search bar; for example, we can find the documentation for PostgreSQL in `https://docs.rs/sqlx/0.5.7/sqlx/postgres/index.html`. On that page, we can see there are `types` modules that we can look at.

6. After we write the content of the migration file, we can run the migration using the following command:

    `sqlx migrate run`

7. After the migration, check the generated database table and see whether the table schema is correct or not. Let's insert the data from the previous chapter. Also, feel free to fill the table with sample data of your choice.

8. After migration, include the `sqlx` crate in our `Cargo.toml` file. We should also include the `uuid` crate as we're going to use PostgreSQL's `uuid` type. Take a look at the API documentation of the crate if you want to enable another RDBMS:

```
sqlx = {version = "0.5.7", features = ["postgres",
"uuid", "runtime-tokio-rustls"]}
uuid = "0.8.2"
```

9. We can delete `lazy_static` from `Cargo.toml` and remove references of `lazy_static!`, `USERS`, and `HashMap` from the `src/main.rs` file. We don't need those, and we are only going to retrieve the `User` data from the database that we inserted earlier. Use the following `SQL INSERT` syntax to insert the previous user data:

    ```
    INSERT INTO public.users
    (uuid, name, age, grade, active)
    VALUES('3e3dd4ae-3c37-40c6-aa64-7061f284ce28'::uuid,
    'John Doe', 18, 1, true);
    ```

10. Modify the `User` struct to follow the database that we've created:

    ```
    use sqlx::FromRow;
    use uuid::Uuid;

    ...

    #[derive(Debug, FromRow)]
    struct User {
        uuid: Uuid,
        name: String,
        age: i16,
        grade: i16,
        active: bool,
    }
    ```

    When `sqlx` retrieves the result from the database, it will be stored in `sqlx::Database::Row`. This type can then be converted to any type that implements `sqlx::FromRow`. Luckily, we can derive `FromRow` as long as all the members implement `sqlx::Decode`. There are a few exceptions that we can use to override `FromRow`. Here is an example:

    ```
    #[derive(Debug, FromRow)]
    #[sqlx(rename_all = "camelCase")]
    struct User {
        uuid: Uuid,
        name: String,
        age: i16,
        grade: i16,
        #[sqlx(rename = "active")]
        present: bool,
        #[sqlx(default)]
    ```

```
    not_in_database: String,
}
```

For `rename_all`, we can use these options: `snake_case`, `lowercase`, `UPPERCASE`, `camelCase`, `PascalCase`, `SCREAMING_SNAKE_CASE`, and `kebab-case`.

`rename` is used when we have different column names and type member names. If a member has no column in the database, and that type has the implementation of the `std::default::Default` trait, we can use the `default` directive.

Why do we use `i16`? The answer is the PostgreSQL type has no mapping to the Rust `u8` type. We can either use `i8`, use a bigger `i16` type, or try implementing `Decode` for `u8`. In this case, we choose to use the `i16` type.

We want the program to read the connection information (`DATABASE_URL`) from the environment variable. In *Chapter 2, Building Our First Rocket Web Application*, we learned how to configure Rocket using standard configuration, but this time, we want to add extra configuration. We can start by adding `serde` to application dependencies in `Cargo.toml`.

`serde` is one of the most used and important libraries in Rust. The name comes from *serialization and deserialization*. It is used for anything that involves serialization and deserialization. It can be used to convert Rust type instances to bytes representations and vice versa, to JSON and vice versa, to YAML, and any other type, as long as they implement the `serde` traits. It can also be used to transcode one type that implements `serde` traits to another type that implements `serde` traits.

If you want to look at the `serde` documentation, you can find it on their website at `https://serde.rs`.

The `serde` documentation mentions many native or third-party supports for many data formats such as JSON, Bincode, CBOR, YAML, MessagePack, TOML, Pickle, RON, BSON, Avro, JSON5, Postcard, URL query strings, Envy, Envy Store, S-expressions, D-Bus's binary wire format, and FlexBuffers.

Let's add the following lines into `Cargo.toml` to include `serde` in our application:

```
[dependencies]
. . .
serde = "1.0.130"
```

11. After that, create a struct that will be used to contain our custom configuration:

```
use serde::Deserialize;

...

#[derive(Deserialize)]
struct Config {
    database_url: String,
}
```

serde already provides the Deserialize macro that can be used in the derive attribute. So far, we have used a lot of macros providing libraries that can be used in the derive attribute, such as Debug, FromRow, Deserialize. The macro system is one of the important Rust features.

12. Implement the routine to read the configuration and map it into the rocket() function:

```
fn rocket() -> Rocket<Build> {
    let our_rocket = rocket::build();

    let config: Config = our_rocket
        .figment()
        .extract()
        .expect("Incorrect Rocket.toml
         configuration");

    ...

    our_rocket
        .manage(visitor_counter)

    ...

}
```

13. Now that the application can get the DATABASE_URL information from environment variables, it's time to initialize the database connection pool and tell Rocket to manage it. Write the following lines:

```
use sqlx::postgres::PgPoolOptions;

...

async fn rocket() -> Rocket<Build> {
```

```
        . . .
        let config: Config = rocket_frame
            .figment()
            .extract()
            .expect("Incorrect Rocket.toml
            configuration");

        let pool = PgPoolOptions::new()
            .max_connections(5)
            .connect(&config.database_url)
            .await
            .expect("Failed to connect to database");
        . . .
        rocket_frame
            .manage(visitor_counter)
            .manage(pool)
        . . .
    }
```

We initialize the connection pool using `PgPoolOptions`. Other databases can use their corresponding type, such as `sqlx::mysql::MySqlPoolOptions` or `sqlx::sqlite::SqlitePoolOptions`. The `connect()` method is an `async` method, so we must make `rocket()` async as well to be able to use the result.

After that, inside the `rocket()` function, we tell Rocket to manage the connection pool.

14. Before using the database connection, we used `lazy_static` and created user objects as references to the USERS hash map. Now, we will use the data from the database, so we need to use concrete objects instead of references. Remove the ampersand (&) from the `Responder` implementation for the User and NewUser structs:

```
impl<'r> Responder<'r, 'r> for User { ... }
struct NewUser(Vec<User>);
impl<'r> Responder<'r, 'r> for NewUser { ... }
```

15. Now, it's time to implement the `user()` function to use the database connection pool to query from the database. Modify the `user()` function as follows:

```
async fn user(
    counter: &State<VisitorCounter>,
    pool: &rocket::State<PgPool>,
    uuid: &str,
) -> Result<User, Status> {
    ...
    let parsed_uuid = Uuid::parse_str(uuid)
    .map_err(|_| Status::BadRequest)?;

    let user = sqlx::query_as!(
        User,
        "SELECT * FROM users WHERE uuid = $1",
        parsed_uuid
    )
    .fetch_one(pool.inner())
    .await;
    user.map_err(|_| Status::NotFound)
}
```

We included the connection pool managed state in the function arguments. After that, we parsed the UUID `&str` parameter into the `Uuid` instance. If there's an error parsing the `uuid` parameter, we change the error to `Status::BadRequest` and return the error.

We then use the `query_as!` macro to send a query to the database server and convert the result to a `User` instance. There are many `sqlx` macros we can use, such as `query!`, `query_file!`, `query_as_unchecked!`, and `query_file_as!`. You can find the documentation for those macros in the `sqlx` API documentation that we mentioned earlier.

The format to use this macro is as follows: `query_as!(RustType, "prepared statement", bind parameter1, ...)`. If you don't need to get the result as a Rust type, you can use the `query!` macro instead.

We then use the `fetch_one()` method. If you want to execute instead of query, for example, to update or delete rows, you can use the `execute()` method. If you want to get all the results, you can use the `fetch_all()` method. You can find other methods to use and their documentation in the `sqlx::query::Query` struct documentation.

We can either keep the `user()` function return as `Option<User>` and use `user.ok()`, or we change the return to `Status::SomeStatus`. Since we change the return type to either `Ok(user)` or `Err(some_error)`, we can just return the `Ok(user)` variant, but we want to use `map_err(|_| Status::NotFound)` to change the error to a `Status` type.

You might be thinking, if we send raw SQL queries to the server, is it possible to do a SQL injection attack? Is it possible to mistakenly get any user input and execute `sqlx::query_as::<_, User>("SELECT * FROM users WHERE name = ?").bind("Name'; DROP TABLE users").fetch_one(pool.inner())?`

The answer is no. `sqlx` prepared and cached each statement. As the result of using a prepared statement, it's more secure than a regular SQL query, and the type returned is also what we expect from the RDBMS server.

16. Let's also change the `users()` function. Like the `user()` function, we want the function to be `async` and get the connection pool from the Rocket managed state. We also want to remove the lifetime from `NewUser` as we are not referencing USERS anymore:

```
async fn users(
    counter: &State<VisitorCounter>,
    pool: &rocket::State<PgPool>,
    name_grade: NameGrade<'_>,
    filters: Option<Filters>,
) -> Result<NewUser, Status> {…}
```

17. After that, we can prepare the prepared statement. We append more conditions to WHERE if the client sends the `filters` request. For PostgreSQL, the prepared statement uses $1, $2, and so on, but for other RDBMSs, you can use ? for the prepared statement:

```
...
let mut query_str = String::from("SELECT * FROM users
WHERE name LIKE $1 AND grade = $2");
if filters.is_some() {
```

```
query_str.push_str(" AND age = $3 AND active =
$4");
}
```

18. Next, write the code to execute the query, but the number of bound parameters may change depending on whether filters exist or not; we use the `query_as` function instead so we can use the `if` branching. We also add `%name%` for the name-bound parameter because we use the `LIKE` operator in the SQL statement. We also have to cast the `u8` type to the `i16` type. And, finally, we use the `fetch_all` method to retrieve all the results. The nice thing with the `query_as!` macro and query as function is they both returned `Vec<T>` or not depending on the `fetch_one` or `fetch_all`:

```
...
let mut query = sqlx::query_as::<_, User>(&query_str)
    .bind(format!("%{}%", &name_grade.name))
    .bind(name_grade.grade as i16);
if let Some(fts) = &filters {
    query = query.bind(fts.age as i16).bind(fts.
    active);
}
let unwrapped_users = query.fetch_all(pool.inner()).
await;
let users: Vec<User> = unwrapped_users.map_err(|_|
Status::InternalServerError)?;
```

19. We can return the result as usual:

```
...
if users.is_empty() {
    Err(Status::NotFound)
} else {
    Ok(NewUser(users))
}
```

Now, let's try calling the user() and users() endpoints again. It should work as it did when we used HashMap. Since we didn't modify the connection options after we wrote connect() on the connection pool, the SQL output is written on the terminal:

```
/* SQLx ping */; rows: 0, elapsed: 944.711µs
SELECT * FROM users …; rows: 1, elapsed: 11.187ms


SELECT
  *
FROM
  users
WHERE
  uuid = $1
```

Here is some more of the output:

```
/* SQLx ping */; rows: 0, elapsed: 524.114µs
SELECT * FROM users …; rows: 1, elapsed: 2.435ms


SELECT
  *
FROM
  users
WHERE
  name LIKE $1
  AND grade = $2
```

In this book, we are not going to use ORM; instead, we are going to use sqlx only, as it is enough for the scope of this book. If you want to use ORM in your application, you can use ORM and query builders from https://github.com/NyxCode/ormx or https://www.sea-ql.org/SeaORM/.

Now that we have learned about State and how to use databases using State, it is time to learn about another Rocket middleware capability, attaching fairings.

Attaching Rocket fairings

In real life, a rocket fairing is a nose cone used to protect the rocket payload. In the Rocket framework, a fairing is not used to protect the payload but is instead used to hook in to any part of the request life cycle and rewrite the payload. Fairings are analogous to middleware in other web frameworks but with few differences.

Other framework middleware may be able to inject any arbitrary data. In Rocket, the fairing can be used to modify the request but cannot be used to add information that is not part of the request. For example, we can use fairings to add a new HTTP header in the requests or responses.

Some web frameworks might be able to terminate and directly respond to incoming requests, but in Rocket, the fairings cannot stop the incoming requests directly; the request must go through the route handling function, and then the route can create the proper response.

We can create a fairing by implementing `rocket::fairing::Fairing` for a type. Let's first see the signature of the trait:

```
#[crate::async_trait]
pub trait Fairing: Send + Sync + Any + 'static {
    fn info(&self) -> Info;
    async fn on_ignite(&self, rocket: Rocket<Build>) ->
    Result { Ok(rocket) }
    async fn on_liftoff(&self, _rocket: &Rocket<Orbit>) { }
    async fn on_request(&self, _req: &mut Request<'_>,
    _data: &mut Data<'_>) {}
    async fn on_response<'r>(&self, _req: &'r Request<'_>,
    _res: &mut Response<'r>) {}
}
```

There are a couple of types we are not familiar with, such as `Build` and `Orbit`. These types are related to phases in Rocket.

Rocket phases

The types that we want to discuss are `Build` and `Orbit`, with the full module paths of `rocket::Orbit` and `rocket::Build`. What are these types? The signature for a Rocket instance is `Rocket<P: Phase>`, which means any P type that implements `rocket::Phase`.

Phase is a **sealed trait**, which means no other type outside the crate can implement this trait. We can define a sealed trait as follows: `pub trait SomeTrait: private::Sealed {}`.

The `Phase` trait is sealed because Rocket authors intended only three phases in the Rocket application: `rocket::Build`, `rocket::Ignite`, and `rocket::Orbit`.

We initialize a Rocket instance through `rocket::build()`, which uses the `Config::figment()` default, or `rocket::custom<T: Provider>(provider: T)`, which uses the custom configuration provider. In this phase, we can also chain the generated instance with custom configuration using `configure<T: Provider>(self, provider: T)`. We can then add a route using `mount()`, register a catcher using `register()`, manage the states using `manage()`, and attach fairings using `attach()`.

After that, we can change the Rocket phase to `Ignite` through the `ignite()` method. In this phase, we have a Rocket instance with the final configuration. We can then send the Rocket to the `Orbit` phase through the `launch()` method or return `Rocket<Build>` and use the `#[launch]` attribute. We can also skip the `Ignite` phase and use `launch()` directly after `build()`.

Let's recall the code that we have created up to now:

```
#[launch]
async fn rocket() -> Rocket<Build> {
    let our_rocket = rocket::build();

    ...

    our_rocket
        .manage(visitor_counter)
        .manage(pool)
        .mount("/", routes![user, users, favicon])
        .register("/", catchers![not_found, forbidden])
}
```

This function generates `Rocket<Build>`, and the `#[launch]` attribute generates the code that uses `launch()`.

The conclusion for this subsection is that the Rocket phase goes from `Build` to `Ignite` to `Launch`. How are those phases related to fairing? Let's discuss this in the next subsection.

Fairing callbacks

Any type implementing fairings must implement one mandatory function, `info()`, which returns `rocket::fairing::Info`. The `Info` struct is defined as follows:

```
pub struct Info {
    pub name: &'static str,
    pub kind: Kind,
}
```

And, `rocket::fairing::Kind` is defined as just an empty struct, `pub struct Kind(_);`, but `Kind` has the **associated constants** of `Kind::Ignite`, `Kind::Liftoff`, `Kind::Request`, `Kind::Response`, and `Kind::Singleton`.

What are associated constants? In Rust, we can declare **associated items**, which are items declared in traits or defined in implementations. For example, we have this piece of code:

```
struct Something {
    item: u8
}
impl Something {
    fn new() -> Something {
        Something {
            item: 8,
        }
    }
}
```

We can use the `Something::new()` **associated function** to create a new instance of the type. There are also **associated methods**, just like associated functions but with `self` as the first parameter. We have already implemented an associated method a couple of times.

We can also define an **associated type** as follows:

```
trait SuperTrait {
    type Super;
}
struct Something;
struct Some;
impl SuperTrait for Something {
```

```
    type Super = Some;
}
```

And, finally, we can have an **associated constant**. Let's take a look at how we can define an associated constant by looking at an example, the simplified source of `rocket::fairing::Kind`:

```
pub struct Kind(usize);
impl Kind {
    pub const Ignite: Kind = Kind(1 << 0);
    pub const Liftoff: Kind = Kind(1 << 1);
    pub const Request: Kind = Kind(1 << 2);
    pub const Response: Kind = Kind(1 << 3);
    pub const Singleton: Kind = Kind(1 << 4);
    ...
}
```

Let's go back to `Info`. We can make an `Info` instance as follows:

```
Info {
    name: "Request Response Tracker",
    kind: Kind::Request | Kind::Response,
}
```

We are saying the value for `kind` is the result of the `OR` bitwise operation between the `kind` associated constants. `Kind::Request` is `1<<2`, which means `100` in binary or `4` in decimal. `Kind::Response` is `1<<3`, which means `1000` in binary or `8` in decimal. The result of `0100 | 1000` is `1100` in binary or `12` in decimal. With this knowledge, we can set the value for the `Info` instance, `kind`, from `00000` to `11111`.

Setting configuration using bitwise is a very common design pattern for packing multiple values into one variable. Some other languages even make this design pattern into its own type and call it **bitset**.

In a type that implements the `Fairing` trait, the mandatory method implementation is `info()`, which returns the `Info` instance. We also have to implement `on_ignite()`, `on_liftoff()`, `on_request()`, and `on_response()` depending on the `kind` instance that we defined. In our case, this means we have to implement `on_request()` and `on_response()`.

Rocket executes our fairing method on different occasions. If we have `on_ignite()`, it will be executed before launch. This type of fairing is special as `on_ignite()` returns `Result`, and if the returned variant is `Err`, it can abort the launch.

For `on_liftoff()`, this method will be executed after launch, which means when Rocket is in the `Orbit` phase.

If we have `on_request()`, it will be executed after Rocket gets the request but before the request is routed. This method will have access to `Request` and `Data`, which means we can modify these two items.

And, `on_response()` will be executed when the route handler has created the response but before the response is sent to the HTTP client. This callback has access to the `Request` and `Response` instances.

`Kind::Singleton` is special. We can create multiple instances of fairings of the same type and attach them to Rocket. But, maybe we only want to allow one instance of the `Fairing` implementing type to be added. We can use `Kind::Singleton` and it will make sure only the last attached instance of this type will be added.

Now that we know more about Rocket phases and `Fairing` callbacks, let's implement the `Fairing` trait in the next subsection.

Implementing and attaching fairings

Right now, our Rocket application manages `VisitorCounter`, but we did not add `State<VisitorCounter>` to the `favicon()` function. We might also want to add new route handling functions, but adding `State<VisitorCounter>` as an argument parameter for every route handling function is cumbersome.

We can change `VisitorCounter` from a managed state into a fairing. At the same time, let's imagine that we have another requirement in our application. We want to have a custom header for the request and response for internal logging purposes. We can do it by adding another fairing to change the incoming requests and responses.

First, let's organize our module usage a little bit. We need to add the fairing-related modules, `rocket::http::Header`, `rocket::Build`, and `rocket::Orbit`, so we can use those for our `VisitorCounter` fairing and another fairing to modify the requests and responses:

```
use rocket::fairing::{self, Fairing, Info, Kind};
use rocket::fs::{relative, NamedFile};
use rocket::http::{ContentType, Header, Status};
```

```
use rocket::request::{FromParam, Request};
use rocket::response::{self, Responder, Response};
use rocket::{Build, Data, Orbit, Rocket, State};
```

Add the `Fairing` trait implementation for `VisitorCounter`. We need to decorate the `impl` with `#[rocket::async_trait]`, since this trait is an `async` trait:

```
#[rocket::async_trait]
impl Fairing for VisitorCounter {
    fn info(&self) -> Info {
        Info {
            name: "Visitor Counter",
            kind: Kind::Ignite | Kind::Liftoff | Kind::
            Request,
        }
    }
}
```

We added the `info()` mandatory method, which returned the `Info` instance. Inside the `Info` instance, we only really need `Kind::Request` as we only need to increment the visitor counter for every incoming request. But this time, we also added `Kind::Ignite` and `Kind::Liftoff` because we want to see when the callback is executed.

Then, we can add the callbacks inside the `impl Fairing` block:

```
async fn on_ignite(&self, rocket: Rocket<Build>) ->
fairing::Result {
    println!("Setting up visitor counter");
    Ok(rocket)
}
async fn on_liftoff(&self, _: &Rocket<Orbit>) {
    println!("Finish setting up visitor counter");
}
async fn on_request(&self, _: &mut Request<'_>, _: &mut
Data<'_>) {
    self.increment_counter();
}
```

What is the return type on the `on_ignite()` method? `rocket::fairing::Result` is defined as `Result<T = Rocket<Build>, E = Rocket<Build>> = Result<T, E>`. This method is used to control whether the program continues or not. For example, we can check the connection to a third-party server to ensure its readiness. If the third-party server is ready to accept connections, we can return `Ok(rocket)`. But, if the third-party server is not available, we can return `Err(rocket)` to halt the launch of Rocket. Notice that `on_liftoff()`, `on_request()`, and `on_response()` do not have a return type, as `Fairing` is designed to only fail when we build Rocket.

For `on_liftoff()`, we just want to print something to the application output. For `on_request()`, we undertake the real purpose of this fairing: increase the counter for every request.

After implementing the `Fairing` trait, we can remove `counter: &State<VisitorCounter>` from the `user()` and `users()` function arguments. We also need to remove `counter.increment_counter();` from the body of those functions.

After we have modified the `user()` and `users()` functions, we can attach the fairing to the Rocket application. Change `manage(visitor_counter)` to `attach(visitor_counter)` in the Rocket initialization code.

Time to see the fairing in action! First, take a look at the initialization sequence. You can see `on_ignite()` is executed in the beginning, and `on_liftoff()` is executed after everything is ready:

```
> cargo run
...
Setting up visitor counter
🔧 Configured for debug.
...
📡 Fairings:
   >> Visitor Counter (ignite, liftoff, request)
...
Finish setting up visitor counter
🚀 Rocket has launched from http://127.0.0.1:8000
```

After that, try calling our route handling function again to see the counter increase again:

```
> curl http://127.0.0.1:8000/user/3e3dd4ae-3c37-40c6-aa64-
7061f284ce28
```

And, in the Rocket output, we can see it increase when we use it as a state:

The number of visitor is: 2

Now, let's implement our second use case, injecting a tracing ID to our requests and responses.

First, modify `Cargo.toml` to ensure the `uuid` crate can generate a random UUID:

```
uuid = {version = "0.8.2", features = ["v4"]}
```

After that, inside `src/main.rs`, we can define the header name we want to inject and a type that works as the fairing:

```
const X_TRACE_ID: &str = "X-TRACE-ID";
struct XTraceId {}
```

Afterward, we can implement the `Fairing` trait for `XtraceId`. This time, we want to have `on_request()` and `on_response()` callbacks:

```
#[rocket::async_trait]
impl Fairing for XTraceId {
    fn info(&self) -> Info {
        Info {
            name: "X-TRACE-ID Injector",
            kind: Kind::Request | Kind::Response,
        }
    }
}
```

Now, write the `on_request()` and `on_response()` implementations inside the `impl Fairing` block:

```
async fn on_request(&self, req: &mut Request<'_>, _: &mut
Data<'_>) {
    let header = Header::new(X_TRACE_ID,
    Uuid::new_v4().to_hyphenated().to_string());
    req.add_header(header);
}
async fn on_response<'r>(&self, req: &'r Request<'_>, res: &mut
Response<'r>) {
```

```
let header = req.headers().get_one(X_TRACE_ID).
unwrap();
res.set_header(Header::new(X_TRACE_ID, header));
}
```

In on_request(), we generate a random UUID and inject the resulting string as one of the request headers. In on_response(), we inject the response with the same header from the request.

Don't forget to initialize and attach this new fairing to the Rocket build and launch process:

```
let x_trace_id = XTraceId {};
our_rocket

    . . .

    .attach(visitor_counter)
    .attach(x_trace_id)

    . . .
```

Rerun the application. We should have a new fairing in the application output and "x-trace-id" in the HTTP response:

```
. . .
 Fairings:
    >> X-TRACE-ID Injector (request, response)
    >> Visitor Counter (ignite, liftoff, request)
. . .
```

Here is another example:

```
curl -v http://127.0.0.1:8000/user/3e3dd4ae-3c37-40c6-aa64-
7061f284ce28
. . .
< x-trace-id: 28c0d523-13cc-4132-ab0a-3bb9ae6153a9
. . .
```

Please note that we can use both State and Fairing in our application. Only use Fairing if we need to call this for every request.

Previously, we created a connection pool and told Rocket to manage it using managed state but Rocket already has a way to connect to the database via its built-in database connection, `rocket_db_pools`, which is a type of fairings. Let's see how we can do it in the next part.

Connecting to a database using rocket_db_pools

Rocket provided a *sanctioned* way to connect to some RDBMSs by using `rocket_db_pools`. That crate provides the database driver integration for Rocket. We are going to learn how to use this crate for connecting to the database. Let's change the connection pool that we made previously from using state into using fairings:

1. We don't need `serde`, as `rocket_db_pools` already has its own configuration. Remove `serde` from `Cargo.toml` and add `rocket_db_pools` as a dependency:

    ```
    [dependencies]
    rocket = "0.5.0-rc.1"
    rocket_db_pools = {version = "0.5.0-rc.1", features =
    ["sqlx_postgres"]}
    ...
    ```

    You can also use different features such as `sqlx_mysql`, `sqlx_sqlite`, `sqlx_mssql`, `deadpool_postgres`, `deadpool_redis`, or `mongodb`.

2. In `Rocket.toml`, remove the line containing the `database_url` configuration and replace it with these lines:

    ```
    [debug.databases.main_connection]
    url = "postgres://username:password@localhost/rocket"
    ```

    You can use `default.databases.main_connection` if you like, and you can also change `main_connection` to whatever name you see fit.

3. In the Cargo library project, we can re-export something in `our_library` using the `pub use something;` syntax, and another library can then use that through `our_library::something`. Remove these `use sqlx...` and `use serde...` lines, as `rocket_db_pools` already re-exported `sqlx` and we don't need `serde` anymore:

    ```
    use serde::Deserialize;
    ...
    use sqlx::postgres::{PgPool, PgPoolOptions};
    use sqlx::FromRow;
    ```

4. Add the following lines to use `rocket_db_pools`. Notice that we can multiline the `use` declaration in the code:

```
use rocket_db_pools::{
    sqlx,
    sqlx::{FromRow, PgPool},
    Connection, Database,
};
```

5. Delete the struct `Config` declaration and add the following lines to declare the database connection type:

```
#[derive(Database)]
#[database("main_connection")]
struct DBConnection(PgPool);
```

The database derives an automatically generated `rocket_db_pools::Database` implementation for the `DBConnection` type. Notice that we wrote the connection name `"main_connection"`, just like what we have set in `Rocket.toml`.

6. Remove the config and connection pool initializations in the `rocket()` function:

```
let config: Config = our_rocket
    .figment()
    .extract()
    .expect("Incorrect Rocket.toml configuration");

let pool = PgPoolOptions::new()
    .max_connections(5)
    .connect(&config.database_url)
    .await
    .expect("Failed to connect to database");
```

7. Add `DBConnection::init()` inside the `rocket()` function and attach it to Rocket:

```
async fn rocket() -> Rocket<Build> {
    ...
    rocket::build()
        .attach(DBConnection::init())
```

```
                    .attach(visitor_counter)
            . . .

    }
```

8. Change the user() and users() functions to use the rocket_db_
    pools::Connection request guard:

    ```
    async fn user(mut db: Connection<DBConnection>, uuid:
    &str) -> Result<User, Status> {

        . . .

        let user = sqlx::query_as!(User, "SELECT * FROM
        users WHERE uuid = $1", parsed_uuid)
            .fetch_one(&mut *db)
            .await;

        . . .

    }
    . . .
    #[get("/users/<name_grade>?<filters..>")]
    async fn users(
        mut db: Connection<DBConnection>,

        . . .
    ) -> Result<NewUser, Status> {

        . . .

        let unwrapped_users = query.fetch_all(&mut
        *db).await;

        . . .

    }
    ```

The application should work just like when we managed the connection pool using state, but with minor differences. Here's the output we see:

📜 **Fairings:**

    . . .
    >> 'main_connection' Database Pool (ignite)

We can see there's a new fairing in the application output but there is no prepared SQL statement in the application output.

Summary

In this chapter, we learned about two Rocket components, `State` and `Fairing`. We can manage state objects and attach fairings upon building rockets, use the `state` objects in route handling functions, and use the `fairing` functions to execute callbacks on the build, after launch, on request, and on response.

We also created counter states and used them in the route handling functions. We also learned how to use `sqlx`, made a database migration, made a database connection pool state, and used `state` to query the database.

Afterward, we learned more about the Rocket initialization process and the building, igniting, and launching phases.

Finally, we changed the counter state into a fairing and created a new fairing to inject a custom HTTP header into the incoming requests and outgoing responses.

Armed with that knowledge, you can create reusable objects between route handling functions, and create a method that can be executed globally between requests and responses.

Our `src/main.rs` file is getting bigger and more complicated; we will learn how to manage our Rust code in modules and plan a more complex application in the next chapter.

5

Designing a User-Generated Application

We are going to write a Rocket application in order to learn more about the Rocket web framework. In this chapter, we are going to design the application and create an application skeleton. Then, we are going to split the application skeleton into smaller manageable modules.

After reading this chapter, you will be able to design and create an application skeleton and modularize your own application to your liking.

In this chapter, we're going to cover the following main topics:

- Designing a user-generated web application
- Planning the user struct
- Creating application routes
- Modularizing a Rocket application

Technical requirements

For this chapter, we have the same technical requirements as the previous chapter. We need a Rust compiler, a text editor, an HTTP client, and a PostgreSQL database server.

You can find the source code for this chapter at `https://github.com/PacktPublishing/Rust-Web-Development-with-Rocket/tree/main/Chapter05`.

Designing a user-generated web application

Up to now, we have gained some fundamental knowledge about the Rocket framework, such as routes, requests, responses, states, and fairings. Let's expand on that knowledge and learn more about the Rocket framework's other capabilities, such as request guards, cookies systems, forms, uploading, and templating, by creating a full-fledged application.

The idea for our application is one that handles various operations for the user, and each user can create and delete user-generated content such as text, photos, or videos.

We can start by creating requirements for what we want to do. In various development methodologies, there are many forms and names for defining requirements, such as user stories, use cases, software requirements, or software requirement specifications.

After specifying the requirements, we can usually create an application skeleton. We can then implement the application and test the implementation.

In our case, because we want to be practical and understand what is going on at the code level, we will specify the requirements and create the application skeleton in the same step.

Let's start by creating a new application. Then, name the application `"our_application"` and include the `rocket` and `rocket_db_pools` crates in `Cargo.toml`:

```
[package]
edition = "2018"
name = "our_application"
version = "0.1.0"

[dependencies]
rocket = {path = "../../../rocket/core/lib/", features =
["uuid"]}
rocket_db_pools = {path = "../../../rocket/contrib/db_pools/
lib/", features =[ "sqlx_postgres"]}
```

Modify the `src/main.rs` file to remove the `main()` function and make sure we have the most basic Rocket application:

```
#[macro_use]
extern crate rocket;

use rocket::{Build, Rocket};

#[launch]
async fn rocket() -> Rocket<Build> {
    rocket::build()
}
```

Let's go to the next step by planning what user data we want to have in our application.

Planning the user struct

Let's write the struct for the user in the application. At the most basic level, we want to have `uuid` with the `Uuid` type as a unique identifier, and `username` with the `String` type as a human-rememberable identifier. Then, we can add extra columns such as `email` and `description` with a `String` type to store a little bit more information about our user.

We also want to have `password` for the user data but having a cleartxt `password` field is not an option. There are a couple of hashing options, but obviously, we cannot use insecure old hashing functions such as `md5` or `sha1`. We can, however, use newer secure hashing encryptions such as `bcrypt`, `scrypt`, or `argon2`. In this book, we will use the `argon2id` function, as it is more resistant to **graphics processing unit** (**GPU**) attacks and side-channel attacks. The password hash can be stored in a text format defined by the **Password Hashing Competition** (**PHC**) string format. As we know the format of the password hash is text, we can use `String` as the `password_hash` type.

We also want a `status` column for our users. The status can be either `active` or `inactive`, so we can use the `bool` type. But, in the future, we might want it to be expandable and have other statuses, such as `confirmed` if we require the user to include email information and confirm their email before they can use our application. We have to use another type.

In Rust, we have `enum`, a type with many variants. We can either have an enum with an **implicit discriminator** or an **explicit discriminator**.

An implicit discriminator enum is an enum in which the member is not given a discriminator; it automatically starts from 0, for example, enum Status {Active, Inactive}. Using an implicit discriminator enum means we have to add a new data type in PostgreSQL using the CREATE TYPE SQL statement, for example, CREATE TYPE status AS ENUM ('active', 'inactive');.

If we use an explicit discriminator enum, that is, an enum in which the member is given a discriminator, we can use the PostgreSQL INTEGER type and map it to rust i32. An explicit discriminator enum looks like the following:

```
enum Status {
    Inactive = 0,
    Active = 1,
}
```

Because it's simpler to use an explicit discriminator enum, we will choose this type for the user status column.

We also want to keep track of when user data is created and when it is updated. The Rust standard library provides std::time for temporal quantification types, but this module is very primitive and not usable for day-to-day operations. There are several attempts to create a good date and time library for Rust, such as the time or chrono crates, and fortunately, sqlx already supports both crates. We chose to use chrono for this book.

Based on those requirements, let's write the struct definition and the sqlx migration for the User type:

1. In Cargo.toml, add the sqlx, chrono, and uuid crates:

    ```
    sqlx = {version = "0.5.9", features = ["postgres",
    "uuid", "runtime-tokio-rustls", "chrono"]}
    chrono = "0.4"
    uuid = {version = "0.8.2", features = ["v4"]}
    ```

2. In src/main.rs, add the UserStatus enum and the User struct:

    ```
    use chrono::{offset::Utc, DateTime};
    use rocket_db_pools::sqlx::FromRow;
    use uuid::Uuid;

    #[derive(sqlx::Type, Debug)]
    #[repr(i32)]
    enum UserStatus {
    ```

```
        Inactive = 0,
        Active = 1,
    }


    #[derive(Debug, FromRow)]
    struct User {
        uuid: Uuid,
        username: String,
        email: String,
        password_hash: String,
        description: String,
        status: UserStatus,
        created_at: DateTime<Utc>,
        updated_at: DateTime<Utc>,
    }
```

Notice that we set the UserStatus enum with an explicit discriminator and, in the User struct, we used UserStatus as the status type.

3. After that, let's set the database URL configuration in the Rocket.toml file:

```
    [default]
    [default.databases.main_connection]
    url = "postgres://username:password@localhost/rocket"
```

4. Afterward, create the database migration using the sqlx migrate add command again, and modify the generated migration file as follows:

```
    CREATE TABLE IF NOT EXISTS users
    (
        uuid          UUID PRIMARY KEY,
        username      VARCHAR NOT NULL UNIQUE,
        email         VARCHAR NOT NULL UNIQUE,
        password_hash VARCHAR NOT NULL,
        description   TEXT,
        status        INTEGER NOT NULL DEFAULT 0,
        created_at    TIMESTAMPTZ NOT NULL DEFAULT
    CURRENT_TIMESTAMP,
        updated_at    TIMESTAMPTZ NOT NULL DEFAULT
```

```
        CURRENT_TIMESTAMP
    );
```

Notice that we set `INTEGER`, which corresponds to `i32` in Rust as a `status` column type. One more thing to notice is because the `UNIQUE` constraints in PostgreSQL already automatically create an index for `username` and `email`, we don't need to add custom indices for those two columns.

Don't forget to run the `sqlx migrate run` command line again to run this migration.

5. Let's initialize the database connection pool fairing by adding these lines in `src/main.rs`:

```
use rocket_db_pools::{sqlx::{FromRow, PgPool}, Database};

...

#[derive(Database)]
#[database("main_connection")]
struct DBConnection(PgPool);

async fn rocket() -> Rocket<Build> {
    rocket::build().attach(DBConnection::init())
}
```

After our `User` struct is ready, the next thing we can do is write the code skeleton for user-related routes, such as creating or deleting users.

Creating user routes

In the previous application, we dealt primarily with getting user data, but in a real-world application, we also want other operations such as inserting, updating, and deleting data. We can expand the two functions to get user data (user and users) into **create, read, update, and delete** (**CRUD**) functions. These four basic functions can be considered fundamental operations of persistent data storage.

In a web application, an architecture style exists to perform operations based on the HTTP method. If we want to get an entity or a collection of entities, we use the HTTP `GET` method. If we want to create an entity, we use the `HTTP POST` method. If we want to update an entity, we use the `HTTP PUT` or `PATCH` method. And finally, if we want to delete an entity, we use the `HTTP DELETE` method. Using those HTTP methods to deliver data uniformly is called **representational state transfer** (**REST**), and an application following that constraint is called **RESTful**.

Before we create RESTful user routes for our application, let's think about what incoming parameters we want to handle and what responses we want to return. In the previous chapters, we have created routes that returned `String`, but most of the web is composed of HTML.

For user route responses, we want HTML, so we can use `rocket::response::content::RawHtml`. We can wrap it in `Result`, with `Status` as the error type. Let's make a type alias to avoid writing `Result<RawHtml<String>, Status>` every time we use it as a route function return type. Add this in `src/main.rs`:

```
type HtmlResponse = Result<RawHtml<String>, Status>;
```

For user route requests, the request payload will be different depending on what the request is. For a function that uses GET to obtain a particular user information, we would need to know the identifier of the user, in our case, it would be `uuid` in the `&str` type. We just need the reference (`&str`) because we are not going to process `uuid`, so we don't need the `String` type:

```
#[get("/users/<_uuid>", format = "text/html")]
async fn get_user(mut _db: Connection<DBConnection>, _uuid:
&str) -> HtmlResponse {
    todo!("will implement later")
}
```

The compiler will emit a warning if we define a variable or pass a parameter but do not use it, so we use an underscore (_) before the variable name in the function argument to suppress the compiler warning for now. We will change the variable to one without an underscore in front of it when we are implementing the function later.

Just like the `unimplemented!` macro, the `todo!` macro is useful for prototyping. The semantic difference is that if we use `todo!`, we are saying that the code will be implemented, but if we use `unimplemented!`, we are not making any promises.

Mount the route and try running the application now and make the HTTP request to this endpoint. You can see how the application will panic, but fortunately, Rocket handles catching the panic in the server using the `std::panic::catch_unwind` function.

For the list of users, we have to think about the scalability of our application. If we have a lot of users, it would not be very efficient if we tried to query all the users. We need to introduce some kind of pagination in our application.

One of the weaknesses of using `Uuid` as an entity ID is that we cannot sort and order the entity by its ID. We have to use another ordered field. Fortunately, we have defined the `created_at` field with `TIMESTAMPZ`, which has a 1-microsecond resolution and can be ordered.

But, be aware that if your application is handling high traffic or will be in distributed systems, the microsecond resolution might not be enough. You can calculate the chance of collision of `TIMESTAMPZ` using a formula to calculate the *birthday paradox*. You can solve this problem with a monotonic ID or hardware and a database supporting a nanosecond resolution, but a highly-scalable application is beyond the scope of this book for the Rocket web framework.

Let's define the `Pagination` struct for now and then we will implement this struct later. As we want to use `Pagination` in the list of users and we will use it as a `request` parameter, we can automatically use `#[derive(FromForm)]` to make auto-generation for the `rocket::form::FromForm` implementation. But, we have to create a new type, `OurDateTime`, because orphan rules mean we cannot implement `rocket::form::FromForField` for `DateTime<Utc>;`:

```
use rocket::form::{self, DataField, FromFormField, ValueField};
...
#[derive(Debug, FromRow)]
struct OurDateTime(DateTime<Utc>);

#[rocket::async_trait]
impl<'r> FromFormField<'r> for OurDateTime {
    fn from_value(_: ValueField<'r>) -> form::Result<'r,
    Self> {
        todo!("will implement later")
    }

    async fn from_data(_: DataField<'r, '_>) -> form::
    Result<'r, Self> {
        todo!("will implement later")
    }
}

#[derive(FromForm)]
struct Pagination {
```

```
    cursor: OurDateTime,
    limit: usize,
}


#[derive(sqlx::Type, Debug, FromFormField)]
#[repr(i32)]
enum UserStatus {

    ...

}


#[derive(Debug, FromRow, FromForm)]
struct User {

    ...

    created_at: OurDateTime,
    updated_at: OurDateTime,
}
```

Now, we can make an unimplemented function for the list of users:

```
#[get("/users?<_pagination>", format = "text/html")]
async fn get_users(mut _db: Connection<DBConnection>, _
pagination: Option<Pagination>) -> HtmlResponse {
    todo!("will implement later")
}
```

We need a page to fill in the form for inputting new user data:

```
#[get("/users/new", format = "text/html")]
async fn new_user(mut _db: Connection<DBConnection>) ->
HtmlResponse {
    todo!("will implement later")
}
```

After that, we can create a function to handle the creation of user data:

```
use rocket::form::{self, DataField, Form, FromFormField,
ValueField};
...
#[post("/users", format = "text/html", data = "<_user>")]
```

```
async fn create_user(mut _db: Connection<DBConnection>, _user:
Form<User>) -> HtmlResponse {
    todo!("will implement later")
}
```

We need a page to modify existing user data:

```
#[get("/users/edit/<_uuid>", format = "text/html")]
async fn edit_user(mut _db: Connection<DBConnection>, _uuid:
&str) -> HtmlResponse {
    todo!("will implement later")
}
```

We need functions to handle updating user data:

```
#[put("/users/<_uuid>", format = "text/html", data = "<_
user>")]
async fn put_user(mut _db: Connection<DBConnection>, _uuid:
&str, _user: Form<User>) -> HtmlResponse {
    todo!("will implement later")
}


#[patch("/users/<_uuid>", format = "text/html", data = "<_
user>")]
async fn patch_user(
    mut _db: Connection<DBConnection>,
    _uuid: &str,
    _user: Form<User>,
) -> HtmlResponse {
    todo!("will implement later")
}
```

What's the difference between PUT and PATCH? Simply put, in REST, a PUT request is used if we want to replace the resource completely, and PATCH is used to update data partially.

The last user-related function is a function to execute HTTP DELETE:

```
#[delete("/users/<_uuid>", format = "text/html")]
async fn delete_user(mut _db: Connection<DBConnection>, _uuid:
&str) -> HtmlResponse {
```

```
        todo!("will implement later")
}
```

After creating user-related route handling functions, we can expand our requirements.

Making user-generated contents

An application that only handles user data is not fun, so we will add the capability for our users to upload and delete posts. Each post can be either a text post, a photo post, or a video post. Let's look at the steps:

1. Create the definition for `Post`:

    ```
    #[derive(sqlx::Type, Debug, FromFormField)]
    #[repr(i32)]
    enum PostType {
        Text = 0,
        Photo = 1,
        Video = 2,
    }


    #[derive(FromForm)]
    struct Post {
        uuid: Uuid,
        user_uuid: Uuid,
        post_type: PostType,
        content: String,
        created_at: OurDateTime,
    }
    ```

    We want to differentiate the type, so we added the `post_type` column. We also want to make a relationship between the user and posts. As we want the user to be able to create many posts, we can create a `user_uuid` field in the struct. The content will be used to store either text content or the file path where we store the uploaded file. We will deal with the data migration on application implementation later.

2. The way each post is presented might be different on our HTML, but it will occupy the same part on the web page, so let's make a `DisplayPostContent` trait and three **newtypes** for each post type, and implement `DisplayPostContent` for each newtype:

```rust
trait DisplayPostContent {
    fn raw_html() -> String;
}


struct TextPost(Post);
impl DisplayPostContent for TextPost {
    fn raw_html() -> String {
        todo!("will implement later")
    }
}


struct PhotoPost(Post);
impl DisplayPostContent for PhotoPost {
    fn raw_html() -> String {
        todo!("will implement later")
    }
}


struct VideoPost(Post);
impl DisplayPostContent for VideoPost {
    fn raw_html() -> String {
        todo!("will implement later")
    }
}
```

3. Finally, we can add the routes for handling `Post`. We can create `get_post`, `get_posts`, `create_post`, and `delete_post`. We also want these routes to be under a user:

```rust
#[get("/users/<_user_uuid>/posts/<_uuid>", format =
"text/html")]
async fn get_post(mut _db: Connection<DBConnection>, _
user_uuid: &str, _uuid: &str) -> HtmlResponse {
    todo!("will implement later")
```

```
}

#[get("/users/<_user_uuid>/posts?<_pagination>", format =
"text/html")]
async fn get_posts(
    mut _db: Connection<DBConnection>,
    _user_uuid: &str,
    _pagination: Option<Pagination>,
) -> HtmlResponse {
    todo!("will implement later")
}


#[post("/users/<_user_uuid>/posts", format = "text/html",
data = "<_upload>")]
async fn create_post(
    mut _db: Connection<DBConnection>,
    _user_uuid: &str,
    _upload: Form<Post>,
) -> HtmlResponse {
    todo!("will implement later")
}


#[delete("/users/<_user_uuid>/posts/<_uuid>", format =
"text/html")]
async fn delete_post(
    mut _db: Connection<DBConnection>,
    _user_uuid: &str,
    _uuid: &str,
) -> HtmlResponse {
    todo!("will implement later")
}
```

After adding post-related types and functions, we can finalize creating the application skeleton in the next subsection.

Finalizing the application

Don't forget to add these routes to the Rocket initialization process:

```
async fn rocket() -> Rocket<Build> {
    rocket::build().attach(DBConnection::init()).mount(
        "/",
        routes![
            get_user,
            get_users,
            new_user,
            create_user,
            edit_user,
            put_user,
            patch_user,
            delete_user,
            get_post,
            get_posts,
            create_post,
            delete_post,
        ],
    )
}
```

We also want to serve the uploaded file through a route:

```
use rocket::fs::{NamedFile, TempFile};
...
#[get("/<_filename>")]
async fn assets(_filename: &str) -> NamedFile {
    todo!("will implement later")
}

async fn rocket() -> Rocket<Build> {
    rocket::build()
        ...
        .mount("/assets", routes![assets])
}
```

Time to add our default error handling! Other frameworks usually have a default error handler for HTTP status codes 404, 422, and 500. Let's make a handler for these codes:

```
use rocket::request::Request;
...
#[catch(404)]
fn not_found(_: &Request) -> RawHtml<String> {
    todo!("will implement later")
}


#[catch(422)]
fn unprocessable_entity(_: &Request) -> RawHtml<String> {
    todo!("will implement later")
}


#[catch(500)]
fn internal_server_error(_: &Request) -> RawHtml<String> {
    todo!("will implement later")
}

async fn rocket() -> Rocket<Build> {
    rocket::build()
        ...
        .register(
            "/",
            catchers![not_found, unprocessable_entity,
            internal_server_error],
        )
}
```

When we run the application using Cargo's run command, the application should launch correctly. But, when we look at src/main.rs, the file has a lot of functions and type definitions. We will modularize our application in the next section.

Modularizing the Rocket application

Remember in *Chapter 1*, *Introducing the Rust Language*, when we made an application with modules? One of the functions of the application source code is to use it as documentation for the people working on the application. A good readable code can be easily further developed and shared with other people on the team.

The compiler does not care whether the program is in one file or multiple files; the resulting application binary is the same. However, people working on a single, long file can get confused very easily.

We are going to split our application source code into smaller files and categorize the files into different modules. Programmers come from various backgrounds and may have their own paradigm on how to split the source code of the application. For example, programmers who are used to writing Java programs may prefer organizing their code based on the logical entities or classes. People who are used to Model-Viev-Controller (MVC) frameworks may prefer putting files in models, views, and controllers folders. People who are used to clean architecture may try to organize their code into layers. But, at the end of the day, what really matters is that the way you organize your code is accepted by the people you work with, and they can all comfortably and easily use the same source code.

Rocket does not have specific guidelines on how to organize the code, but there are two things that we can observe to modularize our application. The first one is the `Cargo` project package layout convention, and the second one is the Rocket parts themselves.

According to Cargo documentation (`https://doc.rust-lang.org/cargo/guide/project-layout.html`), the package layout should be as follows:

```
┌── Cargo.lock
├── Cargo.toml
├── src/
│   ├── lib.rs
│   ├── main.rs
│   └── bin/
│       ├── named-executable.rs
│       ├── another-executable.rs
│       └── multi-file-executable/
│           ├── main.rs
│           └── some_module.rs
├── benches/
│   ├── large-input.rs
│   └── multi-file-bench/
│
```

```
|        ├── main.rs
|        └── bench_module.rs
├── examples/
|   ├── simple.rs
|   └── multi-file-example/
|       ├── main.rs
|       └── ex_module.rs
└── tests/
    ├── some-integration-tests.rs
    └── multi-file-test/
        ├── main.rs
        └── test_module.rs
```

Since we don't have benchmarks, examples, or tests yet, let's focus on the `src` folder. We can split the application into an executable in `src/main.rs` and a library in `src/lib.rs`. It's very common in an executable project to make a small executable code that only calls the library.

We already know Rocket has different parts, so it's a good idea to split the Rocket components into their own module. Let's organize our source code into these files and folders:

```
┌── Cargo.lock
├── Cargo.toml
└── src/
    ├── lib.rs
    ├── main.rs
    ├── catchers
    |   └── put catchers modules here
    ├── fairings
    |   └── put fairings modules here
    ├── models
    |   └── put requests, responses, and database related
modules here
    ├── routes
    |   └── put route handling functions and modules here
    ├── states
    |   └── put states modules here
```

```
├── traits
│       └── put our traits here
└── views
        └── put our templates here
```

1. First, edit the `Cargo.toml` file:

```
[package]
...
[[bin]]
name = "our_application"
path = "src/main.rs"

[lib]
name = "our_application"
path = "src/lib.rs"


[dependencies]
...
```

2. Create the `src/lib.rs` file and the following folders: `src/catchers`, `src/fairings`, `src/models`, `src/routes`, `src/states`, `src/traits`, and `src/views`.

3. After that, create a `mod.rs` file inside each folder: `src/catchers/mod.rs`, `src/fairings/mod.rs`, `src/models/mod.rs`, `src/routes/mod.rs`, `src/states/mod.rs`, `src/traits/mod.rs`, and `src/views/mod.rs`.

4. Then, edit `src/lib.rs`:

```
#[macro_use]
extern crate rocket;

pub mod catchers;
pub mod fairings;
pub mod models;
pub mod routes;
pub mod states;
pub mod traits;
```

5. Write the connection for our database first. Edit src/fairings/mod.rs:

    ```
    pub mod db;
    ```

6. Make a new file, src/fairings/db.rs, and write the file just like the connection we defined earlier in src/main.rs:

    ```
    use rocket_db_pools::{sqlx::PgPool, Database};


    #[derive(Database)]
    #[database("main_connection")]
    pub struct DBConnection(PgPool);
    ```

    Notice that we use only use a smaller number of modules compared to src/main.rs. We also added the pub keyword in order to make the struct accessible from other modules or from src/main.rs.

7. Because the trait is going to be used by the structs, we need to define the trait first. In src/traits/mod.rs, copy the trait from src/main.rs:

    ```
    pub trait DisplayPostContent {
        fn raw_html() -> String;
    }
    ```

8. After that, let's move all of our structs for requests and responses to the src/models folder. Edit src/models/mod.rs as follows:

    ```
    pub mod our_date_time;
    pub mod pagination;
    pub mod photo_post;
    pub mod post;
    pub mod post_type;
    pub mod text_post;
    pub mod user;
    pub mod user_status;
    pub mod video_post;
    ```

9. Then, create the files and copy the definition from src/main.rs to those files. The first one is src/models/our_date_time.rs:

    ```
    use chrono::{offset::Utc, DateTime};
    use rocket::form::{self, DataField, FromFormField,
    ```

```
ValueField};

#[derive(Debug)]
pub struct OurDateTime(DateTime<Utc>);

#[rocket::async_trait]
impl<'r> FromFormField<'r> for OurDateTime {
    fn from_value(_: ValueField<'r>) -> form::
    Result<'r, Self> {
        todo!("will implement later")
    }

    async fn from_data(_: DataField<'r, '_>) ->
    form::Result<'r, Self> {
        todo!("will implement later")
    }
}
```

10. Next is src/models/pagination.rs:

```
use super::our_date_time::OurDateTime;

#[derive(FromForm)]
pub struct Pagination {
    pub next: OurDateTime,
    pub limit: usize,
}
```

Notice the use declaration uses the super keyword. The Rust module is organized by hierarchy, with a module containing other modules. The super keyword is used to access the module containing the current module. The super keyword can be chained, for example, use super::super::SomeModule;.

11. After that, write src/models/post_type.rs:

```
use rocket::form::FromFormField;
use rocket_db_pools::sqlx;

#[derive(sqlx::Type, Debug, FromFormField)]
```

```
#[repr(i32)]
pub enum PostType {
    Text = 0,
    Photo = 1,
    Video = 2,
}
```

12. Also, write src/models/post.rs:

```
use super::our_date_time::OurDateTime;
use super::post_type::PostType;
use rocket::form::FromForm;
use uuid::Uuid;

#[derive(FromForm)]
pub struct Post {
    pub uuid: Uuid,
    pub user_uuid: Uuid,
    pub post_type: PostType,
    pub content: String,
    pub created_at: OurDateTime,
}
```

And then, write src/models/user_status.rs:

```
use rocket::form::FromFormField;
use rocket_db_pools::sqlx;

#[derive(sqlx::Type, Debug, FromFormField)]
#[repr(i32)]
pub enum UserStatus {
    Inactive = 0,
    Active = 1,
}
```

13. Write src/models/user.rs:

```
use super::our_date_time::OurDateTime;
use super::user_status::UserStatus;
use rocket::form::FromForm;
```

```
use rocket_db_pools::sqlx::FromRow;
use uuid::Uuid;

#[derive(Debug, FromRow, FromForm)]
pub struct User {
    pub uuid: Uuid,
    pub username: String,
    pub email: String,
    pub password_hash: String,
    pub description: Option<String>,
    pub status: UserStatus,
    pub created_at: OurDateTime,
    pub updated_at: OurDateTime,
}
```

And then, write the three post newtypes, src/models/photo_post.rs, src/
models/text_post.rs, and src/models/video_post.rs:

```
use super::post::Post;
use crate::traits::DisplayPostContent;

pub struct PhotoPost(Post);

impl DisplayPostContent for PhotoPost {
    fn raw_html() -> String {
        todo!("will implement later")
    }
}
use super::post::Post;
use crate::traits::DisplayPostContent;

pub struct TextPost(Post);

impl DisplayPostContent for TextPost {
    fn raw_html() -> String {
        todo!("will implement later")
    }
}
```

```
use super::post::Post;
use crate::traits::DisplayPostContent;

pub struct VideoPost(Post);

impl DisplayPostContent for VideoPost {
    fn raw_html() -> String {
        todo!("will implement later")
    }
}
```

In all three files, we use the `crate` keyword in the `use` declaration. We have discussed the `super` keyword before; the `crate` keyword is referring to the current library we are working on, which is the `our_application` library. In Rust 2015 edition, it's written as a double semicolon (`::`), but since Rust 2018 edition, `::` changed to `crate`. Now, `::` means the root path of the external crate, for example, `::rocket::fs::NamedFile;`.

Besides `super`, `::`, and `crate`, there are a couple more **path qualifiers** we can use in a `use` declaration: `self` and `Self`. We can use `self` to avoid ambiguity when referring to items in code, as shown in this example:

```
use super::haha;
mod a {
    fn haha() {}
    fn other_func() {
        self::haha();
    }
}
```

`Self` is used to refer to an associated type in a trait, as shown in this example:

```
trait A {
  type Any;
  fn any(&self) -> Self::Any;
}
struct B;
impl A for B {
  type Any = usize;
  fn any(&self) -> self::Any {
```

```
        100
    }
}
```

14. Now, let's get back to the application skeleton. After all the structs, it's time to write routes for the application. Modify `src/routes/mod.rs`:

```rust
use rocket::fs::NamedFile;
use rocket::http::Status;
use rocket::response::content::RawHtml;

pub mod post;
pub mod user;

type HtmlResponse = Result<RawHtml<String>, Status>;

#[get("/<_filename>")]
pub async fn assets(_filename: &str) -> NamedFile {
    todo!("will implement later")
}
```

We could put the function handling assets in their own Rust file, but since there's only one function and it's very simple, we can just put the function in the `mod.rs` file.

15. Next, create and write `src/routes/post.rs`:

```rust
use super::HtmlResponse;
use crate::fairings::db::DBConnection;
use crate::models::{pagination::Pagination, post::Post};
use rocket::form::Form;
use rocket_db_pools::Connection;

#[get("/users/<_user_uuid>/posts/<_uuid>", format = "text/html")]
pub async fn get_post(
    mut _db: Connection<DBConnection>,
    _user_uuid: &str,
    _uuid: &str,
) -> HtmlResponse {
```

```
        todo!("will implement later")
    }


    #[get("/users/<_user_uuid>/posts?<_pagination>", format =
    "text/html")]
    pub async fn get_posts(
        mut _db: Connection<DBConnection>,
        _user_uuid: &str,
        _pagination: Option<Pagination>,
    ) -> HtmlResponse {
        todo!("will implement later")
    }


    #[post("/users/<_user_uuid>/posts", format = "text/html",
    data = "<_upload>")]
    pub async fn create_post(
        mut _db: Connection<DBConnection>,
        _user_uuid: &str,
        _upload: Form<Post>,
    ) -> HtmlResponse {
        todo!("will implement later")
    }


    #[delete("/users/<_user_uuid>/posts/<_uuid>", format =
    "text/html")]
    pub async fn delete_post(
        mut _db: Connection<DBConnection>,
        _user_uuid: &str,
        _uuid: &str,
    ) -> HtmlResponse {
        todo!("will implement later")
    }
```

16. Create and write src/routes/user.rs:

```
use super::HtmlResponse;
use crate::fairings::db::DBConnection;
use crate::models::{pagination::Pagination, user::User};
```

```
use rocket::form::Form;
use rocket_db_pools::Connection;

#[get("/users/<_uuid>", format = "text/html")]
pub async fn get_user(mut _db: Connection<DBConnection>,
_uuid: &str) -> HtmlResponse {
    todo!("will implement later")
}


#[get("/users?<_pagination>", format = "text/html")]
pub async fn get_users(
    mut _db: Connection<DBConnection>,
    _pagination: Option<Pagination>,
) -> HtmlResponse {
    todo!("will implement later")
}


#[get("/users/new", format = "text/html")]
pub async fn new_user(mut _db: Connection<DBConnection>)
-> HtmlResponse {
    todo!("will implement later")
}


#[post("/users", format = "text/html", data = "<_user>")]
pub async fn create_user(mut _db:
Connection<DBConnection>, _user: Form<User>) ->
HtmlResponse {
    todo!("will implement later")
}


#[get("/users/edit/<_uuid>", format = "text/html")]
pub async fn edit_user(mut _db: Connection<DBConnection>,
_uuid: &str) -> HtmlResponse {
    todo!("will implement later")
}


#[put("/users/<_uuid>", format = "text/html", data = "<_
```

```
user>")]
pub async fn put_user(
    mut _db: Connection<DBConnection>,
    _uuid: &str,
    _user: Form<User>,
) -> HtmlResponse {
    todo!("will implement later")
}


#[patch("/users/<_uuid>", format = "text/html", data =
"<_user>")]
pub async fn patch_user(
    mut _db: Connection<DBConnection>,
    _uuid: &str,
    _user: Form<User>,
) -> HtmlResponse {
    todo!("will implement later")
}


#[delete("/users/<_uuid>", format = "text/html")]
pub async fn delete_user(mut _db:
Connection<DBConnection>, _uuid: &str) -> HtmlResponse {
    todo!("will implement later")
}
```

17. And, to finalize the library, add the catchers in src/catchers/mod.rs:

```
use rocket::request::Request;
use rocket::response::content::RawHtml;


#[catch(404)]
pub fn not_found(_: &Request) -> RawHtml<String> {
    todo!("will implement later")
}


#[catch(422)]
pub fn unprocessable_entity(_: &Request) ->
RawHtml<String> {
```

```
        todo!("will implement later")
}


#[catch(500)]
pub fn internal_server_error(_: &Request) ->
RawHtml<String> {
        todo!("will implement later")
}
```

18. When the library is ready, we can modify `src/main.rs` itself:

```
#[macro_use]
extern crate rocket;

use our_application::catchers;
use our_application::fairings::db::DBConnection;
use our_application::routes::{self, post, user};
use rocket::{Build, Rocket};
use rocket_db_pools::Database;

#[launch]
async fn rocket() -> Rocket<Build> {
    rocket::build()
        .attach(DBConnection::init())
        .mount(
            "/",
            routes![
                user::get_user,
                user::get_users,
                user::new_user,
                user::create_user,
                user::edit_user,
                user::put_user,
                user::patch_user,
                user::delete_user,
                post::get_post,
                post::get_posts,
```

```
                post::create_post,
                post::delete_post,
            ],
        )
        .mount("/assets", routes![routes::assets])
        .register(
            "/",
            catchers![
                catchers::not_found,
                catchers::unprocessable_entity,
                catchers::internal_server_error
            ],
        )
    }
```

Our `src/main.rs` file has become cleaner.

Now, if we want to add more structs or routes, we can easily add new modules in the corresponding folders. We can also add more states or fairings and easily find the file location for those items.

Summary

In this chapter, we learned how to design an application, create a Rocket application skeleton, and organize the Rust application into smaller manageable modules.

We also learned about concepts such as CRUD and RESTful applications, Rust enum discriminators, and Rust path qualifiers.

Hopefully, after reading this chapter, you can apply those concepts to help you organize your code better.

We will start implementing this application and learn more about Rust and Rocket concepts such as templating, request guards, cookies, and JSON, in the following chapters.

Part 2:
An In-Depth Look at Rocket Web Application Development

In this part, you will learn about the intermediate concepts of the Rust language, such as Rust vectors, error handling, Rust Option and Result, enums, looping, matching, closures, and async. You will learn more about Rocket's built-in guards, creating custom request guards, forms, cookies, uploading files, templates, and JSON by using the `serde` crate.

This part comprises the following chapters:

- *Chapter 6, Implementing User CRUD*
- *Chapter 7, Handling Errors in Rust and Rocket*
- *Chapter 8, Serving Static Assets and Templates*
- *Chapter 9, Displaying Users' Post*
- *Chapter 10, Uploading and Processing Posts*
- *Chapter 11, Securing and Adding an API and JSON*

6
Implementing User CRUD

In the previous chapter, we created a rough outline for the application. In this chapter, we are going to implement the endpoints for managing users. By implementing the endpoints in this chapter, you are going to learn about HTTP basic operations for an entity, that is, creating, reading, updating, and deleting an entity.

In addition, you are going to learn how to construct HTML and an HTML form, send the form payload to the server, validate and sanitize the form payload, hash the password payload, and handle failure by redirecting to another endpoint with a message.

Along with implementing the endpoints, you will also learn how to query single and multiple rows from the database, and how to insert, update, and delete a row from the database.

In this chapter, we're going to cover the following main topics:

- Implementing GET user
- Implementing GET users
- Implementing POST user
- Implementing PUT and PATCH user
- Implementing DELETE user

Technical requirements

For this chapter, we have the same technical requirements as the previous chapter. We need a Rust compiler, a text editor, an HTTP client, and a PostgreSQL database server.

You can find the source code for this chapter at `https://github.com/PacktPublishing/Rust-Web-Development-with-Rocket/tree/main/Chapter06`.

Implementing GET user

Let's look at the steps to implement this:

1. We'll start with the basics by implementing the `get_user()` function in `src/routes/user.rs`:

    ```
    #[get("/users/<_uuid>", format = "text/html")]
    pub async fn get_user(mut _db: Connection<DBConnection>,
    _uuid: &str) -> HtmlResponse {
        todo!("will implement later")
    }
    ```

    Before we implement `get_user()`, we want to prepare the other routines that we will use. For example, we want to return HTML, so we need to create a `const` of `&'static str` in the same `src/routes/user.rs` file as our HTML template.

2. We will create two separate instances of `const` so we can insert different contents between the HTML prefix and suffix:

    ```
    const USER_HTML_PREFIX: &str = r#"<!DOCTYPE html>
    <html lang="en">
    <head>
    <meta charset="utf-8" />
    <title>Our Application User</title>
    </head>
    <body>"#;

    const USER_HTML_SUFFIX: &str = r#"</body>
    </html>"#;
    ```

After that, we will create two methods for a `User` struct. The first one is the find method to find the entry in the database server, and the second one is to create an HTML string for a `User` instance.

3. In the `src/models/user.rs` file, add the following lines to the `use` directives:

```
use rocket_db_pools::sqlx::{FromRow, PgConnection};
use std::error::Error;
```

4. Then, create an `impl` block for `User`:

```
impl User{}
```

5. Inside the block, add the `find()` method:

```
pub async fn find(connection: &mut PgConnection, uuid:
&str) -> Result<Self, Box<dyn Error>> {
    let parsed_uuid = Uuid::parse_str(uuid)?;
    let query_str = "SELECT * FROM users WHERE uuid =
    $1";
    Ok(sqlx::query_as::<_, Self>(query_str)
        .bind(parsed_uuid)
        .fetch_one(connection)
        .await?)
}
```

We created a similar method before. The first thing we do is parse the UUID (Universal Unique Identifier) `&str` into a `Uuid` instance and use the question mark operator (?) to quickly return `Box<dyn Error>`. After that, we define the SQL query string, `query_str`, and finally, we return the `User` instance.

One thing a little different here is that we are passing the mutable reference to `PgConnection` itself, instead of a mutable reference to `Connection<DBConnection>`.

Remember previously, we used `Connection<DBConnection>` as in the following:

```
pub async fn find(db: &mut Connection<DBConnection>,
uuid: &str) -> ... {

    ...

        .fetch_one(&mut *db)
        .await?)
}
```

We first dereference db using an asterisk (*) operator. The connection implements `std::ops::Deref`, and its implementation exposes the new type, `DBConnection`, which is a wrapper for `sqlx::PgPool`, an alias to `sqlx::Pool<sqlx::Postgres>`.

The `sqlx::Executor` trait is implemented for `sqlx::PgConnection`, a struct that implements the `sqlx::Connection` trait, representing a single connection to the database. The `sqlx::Executor` trait is also implemented for `&sqlx::Pool`, an asynchronous pool of SQLx database connections.

Since various `sqlx` methods (such as `fetch_all`, `fetch_one`, `fetch_many`, `fetch`, `execute`, and `execute_many`) accept the generic type E, which is bound by the `sqlx::Executor` trait, we can use either the reference to the pool itself or the connection obtained from the pool in those methods.

There's a problem with the `find()` method since `OurDateTime` is a type unknown by `sqlx`.

6. Add the following directive in `src/models/our_date_time.rs`:

```
#[derive(Debug, sqlx::Type)]
#[sqlx(transparent)]
pub struct OurDateTime(DateTime<Utc>);
```

The `transparent` directive automatically generates implementations referring to the implementation of the inner type, which in our case is `DateTime<Utc>`.

7. Beside the `find()` method, let's implement another method to convert `User` to an HTML `String`:

```
pub fn to_html_string(&self) -> String {
    format!(
        r#"<div><span class="label">UUID:
        </span>{uuid}</div>
```

```
<div><span class="label">Username: </span>{username}</
div>
<div><span class="label">Email: </span>{email}</div>
<div><span class="label">Description: </
span>{description}</div>
<div><span class="label">Status: </span>{status}</div>
<div><span class="label">Created At: </span>{created_
at}</div>
<div><span class="label">Updated At: </span>{updated_
at}</div>"#,
        uuid = self.uuid,
        username = self.username,
        email = self.email,
        description = self.description.as_ref().
        unwrap_or(&String::from("")),
        status = self.status.to_string(),
        created_at = self.created_at.0.to_rfc3339(),
        updated_at = self.updated_at.0.to_rfc3339(),
    )
}
```

8. Since the `OurDateTime` member is private but we access it like `self.created_at.0.to_rfc3339()`, it's going to create an error when we compile it. To resolve it, convert the member of `OurDateTime` in `src/models/our_date_time.rs` to public:

    ```
    pub struct OurDateTime(pub DateTime<Utc>);
    ```

    We need to implement the `to_string()` method for `UserStatus` as well. We can choose to implement `to_string()`, or we can implement `std::fmt::Display`, which automatically provides `to_string()`. As a bonus, with the `Display` trait, we can also use it in the `format!("{}", something)` macro.

9. Modify `src/models/user_status.rs` as follows:

    ```
    use std::fmt;
    ...
    impl fmt::Display for UserStatus {
        fn fmt(&self, f: &mut fmt::Formatter<'_>) ->
    ```

```
        fmt::Result {
            match *self {
                UserStatus::Inactive => write!(f,
                "Inactive"),
                UserStatus::Active => write!(f, "Active"),
            }
        }
    }
```

10. It's time to implement the get_user() function in src/routes/user.rs:

```
use rocket::http::Status;
use rocket::response::content::RawHtml;
use rocket_db_pools::{sqlx::Acquire, Connection};


#[get("/users/<uuid>", format = "text/html")]
pub async fn get_user(mut db: Connection<DBConnection>,
uuid: &str) -> HtmlResponse {
    let connection = db
        .acquire()
        .await
        .map_err(|_| Status::InternalServerError)?;
}
```

First, we add the required use directives, and then we remove the underscore from _db and _uuid to mark it as a used variable. Then, we acquire a single connection from the database pool or return InternalServerError if there's something wrong.

11. After we set the connection variable, we can execute the find() method we defined previously:

```
    ...
    let user = User::find(connection, uuid)
        .await
        .map_err(|_| Status::NotFound)?;
```

We only expose a simple error, NotFound, for this case, but a more complex application should handle the error properly, such as logging the error and returning the proper error status and error messages.

12. Finally, we can construct the HTML string and return it:

```
        . . .

        let mut html_string = String::from(USER_HTML_PREFIX);
        html_string.push_str(&user.to_html_string());
        html_string.push_str(format!(r#"<a href="
        /users/edit/{}">Edit User</a>"#,
        user.uuid).as_ref());
        html_string.push_str(r#"<a href="/users">User
        List</a>"#);
        html_string.push_str(USER_HTML_SUFFIX);
        Ok(RawHtml(html_string))
```

Notice we added two links for editing this user and to go to /users.

In the next section, we are going to implement the get_users() function so the application can handle a /users endpoint.

Implementing GET users

Now, let's implement get_users(). Here is a quick reminder of what the function looked like in the previous chapter:

```
#[get("/users?<_pagination>", format = "text/html")]
pub async fn get_users(
    mut _db: Connection<DBConnection>,
    _pagination: Option<Pagination>,
) -> HtmlResponse {
    todo!("will implement later")
}
```

As before, we should prepare the routines we're going to use:

1. In src/models/user.rs, create a method called find_all, as in the following:

```
use super::pagination::{Pagination};
use crate::fairings::db::DBConnection;
use rocket_db_pools::Connection;
use rocket_db_pools::sqlx::{Acquire, FromRow,
PgConnection};
```

```
. . .
impl User {
. . .
    pub async fn find_all(
        db: &mut Connection<DBConnection>,
        pagination: Option<Pagination>,
    ) -> Result<(Vec<Self>, Option<Pagination>),
    Box<dyn Error>> {
        if pagination.is_some() {
            return Self::find_all_with_pagination(db,
            &(pagination.unwrap())).await;
        } else {
            return Self::find_all_without_
            pagination(db).await;
        }
    }
}
```

The parameter for find_all is Connection, which has the connection pool and the optional Pagination.

If the function is successful, we want to return a vector of User and Pagination. We can wrap it as a *tuple* in parentheses (), but there's a possibility that there's no further row in the database, so we wrap the returned Pagination in Option. We then split it into two methods to make it easier to read: find_all_without_ pagination and find_all_with_pagination.

2. Let's modify src/models/pagination.rs a little bit, and add DEFAULT_ LIMIT to limit the number of users we want to fetch at a single time:

```
pub const DEFAULT_LIMIT: usize = 10;
```

3. We can then create and implement the function for the base case in src/models/ user.rs, find_all_without_pagination:

```
use super::pagination::{Pagination, DEFAULT_LIMIT};
. . .
async fn find_all_without_pagination(db: &mut
Connection<DBConnection>) -> Result<(Vec<Self>,
Option<Pagination>), Box<dyn Error>> {
    let query_str = "SELECT * FROM users ORDER BY
```

```
created_at DESC LIMIT $1";
let connection = db.acquire().await?;
let users = sqlx::query_as::<_, Self>(query_str)
    .bind(DEFAULT_LIMIT as i32)
    .fetch_all(connection)
    .await?;
}
```

Like the find() method, we define query_str and execute the query, and bind the Vec<User> result to the users variable. But, why do we pass the &mut Connection<DBConnection> database connection pool this time? Let's continue the function first:

```
{
    ...
    let mut new_pagination: Option<Pagination> = None;
    if users.len() == DEFAULT_LIMIT {
        let query_str = "SELECT EXISTS(SELECT 1 FROM
        users WHERE created_at < $1 ORDER BY
        created_at DESC LIMIT 1)";
        let connection = db.acquire().await?;
        let exists = sqlx::query_as::<_,
        BoolWrapper>(query_str)
            .bind(&users.last().unwrap().created_at)
            .fetch_one(connection)
            .await?;
        if exists.0 {
            new_pagination = Some(Pagination {
                next: users.last().unwrap().
                created_at.to_owned(),
                limit: DEFAULT_LIMIT,
            });
        }
    }
    Ok((users, new_pagination))
}
```

We then prepare the returned pagination, setting it to None first. If the fetched users are equal to DEFAULT_LIMIT, there's a possibility of a next row, so we do the second query to the database. Since we cannot reuse a single connection, we have to obtain a new connection again from the database pool. That's why we pass &mut Connection<DBConnection> to find_all and find_all_without_ pagination instead of &mut PgConnection. If there's the next row, we can return the pagination wrapped in Some(). But, what is BoolWrapper? We need to set a type to put the result of the "SELECT EXISTS..." query.

4. Add pub mod bool_wrapper; to src/models/mod.rs and create a new file, src/models/bool_wrapper.rs, with the following content:

```
use rocket_db_pools::sqlx::FromRow;


#[derive(FromRow)]
pub struct BoolWrapper(pub bool);
```

Don't forget to add use super::bool_wrapper::BoolWrapper; to src/models/user.rs.

5. Now, it's time to implement find_all_with_pagination:

```
async fn find_all_with_pagination(db: &mut
Connection<DBConnection>, pagination: &Pagination) ->
Result<(Vec<Self>, Option<Pagination>), Box<dyn Error>> {
    let query_str =
        "SELECT * FROM users WHERE created_at < $1
        ORDER BY created_at DESC LIMIT
2";
    let connection = db.acquire().await?;
    let users = sqlx::query_as::<_, Self>(query_str)
        .bind(&pagination.next)
        .bind(DEFAULT_LIMIT as i32)
        .fetch_all(connection)
        .await?;
    let mut new_pagination: Option<Pagination> = None;
    if users.len() == DEFAULT_LIMIT {
        let query_str = "SELECT EXISTS(SELECT 1 FROM
        users WHERE created_at < $1 ORDER BY
        created_at DESC LIMIT 1)";
```

```
        let connection = db.acquire().await?;
        let exists = sqlx::query_as::<_,
        BoolWrapper>(query_str)
            .bind(&users.last().unwrap().created_at)
            .fetch_one(connection)
            .await?;
        if exists.0 {
            new_pagination = Some(Pagination {
                next: users.last().unwrap().
                created_at.to_owned(),
                limit: DEFAULT_LIMIT,
            });
        }
    }
    Ok((users, new_pagination))
}
```

The private method works like `find_all_without_pagination`, but we add a `WHERE` condition to start querying from a certain point.

6. Now, it's time to implement the `get_users()` function:

```
#[get("/users?<pagination>", format = "text/html")]
pub async fn get_users(mut db: Connection<DBConnection>,
    pagination: Option<Pagination>) -> HtmlResponse {
    let (users, new_pagination) = User::find_all(&mut
    db, pagination)
        .await
        .map_err(|_| Status::NotFound)?;
}
```

7. After we have obtained `users` and `new_pagination`, we can construct the HTML for the return value:

```
    ...
    let mut html_string = String::from(USER_HTML_PREFIX);
    for user in users.iter() {
        html_string.push_str(&user.to_html_string());
        html_string
```

```
        .push_str(format!(r#"<a href="/users/{}">See
        User</a><br/>"#, user.uuid).as_ref());
    html_string.push_str(
        format!(r#"<a href="/users/edit/{}">Edit
        User</a><br/>"#, user.uuid).as_ref(),
    );
}
```

8. Append the link to the next page if we have `new_pagination`:

```
if let Some(pg) = new_pagination {
    html_string.push_str(
        format!(
            r#"<a href="/users?pagination.next={}&
            pagination.limit={}">Next</a><br/>"#,
            &(pg.next.0).timestamp_nanos(),
            &pg.limit,
        )
        .as_ref(),
    );
}
```

Notice we use `timestamp_nanos()` to convert time to `i64` to make it easier to transport in the HTML.

9. To finalize the function, append the following lines:

```
html_string.push_str(r#"<a href="/users/new">New user</
a>"#);
html_string.push_str(USER_HTML_SUFFIX);
Ok(RawHtml(html_string))
```

10. Now, we have to implement `FromFormField` for `OurDateTime` since we are using `OurDateTime` in the pagination. In `src/models/our_date_time.rs`, add the required use directives:

```
use chrono::{offset::Utc, DateTime, TimeZone};
use rocket::data::ToByteUnit;
```

11. Because we are cloning `OurDateTime` inside the `User` implementation (`users.last().unwrap().created_at.to_owned()`), we also need to derive `Clone` for `OurDateTime`:

```
#[derive(Debug, sqlx::Type, Clone)]
```

12. For the `from_value` implementation, we are just parsing `i64` from the request parameter and converting it to the `OurDateTime` object:

```
impl<'r> FromFormField<'r> for OurDateTime {
    fn from_value(field: ValueField<'r>) -> form::
    Result<'r, Self> {
        let timestamp = field.value.parse::<i64>()?;
        Ok(OurDateTime(
        Utc.timestamp_nanos(timestamp)))
    }
    ...
}
```

13. But, for `from_data`, we have to be more involved because we have to convert the request to `bytes`, convert it again into `&str`, and finally, into `i64`. First, we get the Rocket limit for the form:

```
async fn from_data(field: DataField<'r, '_>) ->
form::Result<'r, Self> {
    let limit = field
        .request
        .limits()
        .get("form")
        .unwrap_or_else(|| 8.kibibytes());
    ...
}
```

14. Then, get the `bytes` from the request:

```
let bytes = field.data.open(limit).into_bytes().await?;
if !bytes.is_complete() {
    return Err((None, Some(limit)).into());
}
let bytes = bytes.into_inner();
```

15. And finally, we convert `bytes` to `&str`, parse it as `i64`, and convert it to `OurDateTime`:

```
let time_string = std::str::from_utf8(&bytes)?;
let timestamp = time_string.parse::<i64>()?;
Ok(OurDateTime(Utc.timestamp_nanos(timestamp)))
```

Now, `get_user` and `get_users` are ready, but we do not have any data yet. In the next section, we are going to implement the `new_user` and `create_user` functions so we can insert user data through an HTML form.

Implementing POST user

To create a user, we are going to use the `new_user` and `create_user` functions. The `new_user()` function is relatively easy to implement; we just need to serve an HTML page with a form for a user to fill in.

Let's look at the steps:

1. Implement the `new_user()` function in `src/routes/user.rs`:

```
#[get("/users/new", format = "text/html")]
pub async fn new_user() -> HtmlResponse {
    let mut html_string = String::from(USER_HTML_
    PREFIX);
    html_string.push_str(
        r#"<form accept-charset="UTF-8" action="/
        users" autocomplete="off" method="POST">
<div>
    <label for="username">Username:</label>
    <input name="username" type="text"/>
</div>
<div>
    <label for="email">Email:</label>
    <input name="email" type="email"/>
</div>
<div>
    <label for="password">Password:</label>
    <input name="password" type="password"/>
</div>
```

```
<div>
    <label for="password_confirmation">Password
    Confirmation:</label>
    <input name="password_confirmation"
    type="password"/>
</div>
<div>
    <label for="description">Tell us a little bit
    more about yourself:</label>
    <textarea name="description"></textarea>
</div>
<button type="submit" value="Submit">Submit</
    button>
</form>"#,
    );
    html_string.push_str(USER_HTML_SUFFIX);
    Ok(RawHtml(html_string))
}
```

In the HTML, we set the form tag action attribute to "/users" and the method attribute to "POST". This corresponds to the create_user route in our application. On the HTML page, we have fields for username, email, password, password_confirmation, and description. We then insert the button to submit and serve the html_string to the client application.

2. Try running the application now and open http://127.0.0.1:8000/users/new in the web browser. Finally, we have something we can render in the browser:

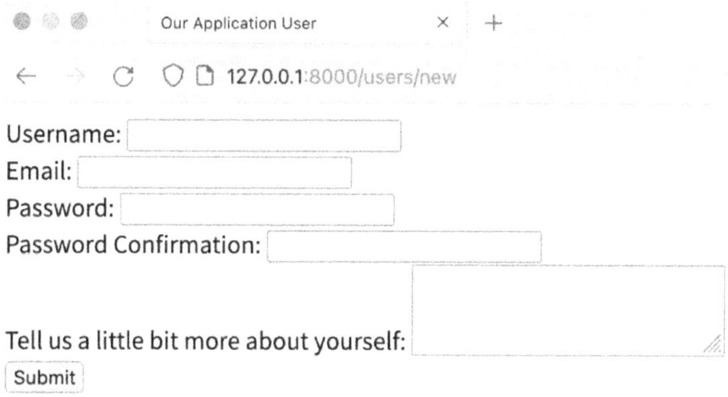

Figure 6.1 – New user page

3. As previously, before implementing the `create_user()` function, we want to
    create other routines first. Since the HTML form has no one-to-one mapping to a
    User struct, we create another struct. Put this struct in `src/models/user.rs`:

```
#[derive(Debug, FromForm)]
pub struct NewUser<'r> {
    #[field(validate = len(5..20).or_else(msg!("name
    cannot be empty")))]
    pub username: &'r str,
    pub email: &'r str,
    pub password: &'r str,
    #[field(validate = eq(self.password).or_
    else(msg!("password confirmation mismatch")))]
    pub password_confirmation: &'r str,
    #[field(default = "")]
    pub description: Option<&'r str>,
}
```

We set the derive `FromForm` trait for `NewUser`, so on top of the struct fields, we
use the `field` directive. This directive can be used to match the request payload
field name to the struct field name, set the default value, and validate the content of
the field.

If we have HTML form field that is different from the struct field name, we can
rename the field using field directive like the following:

```
#[field(name = uncased("html-field-name"))]
```

It can also be done as follows:

```
#[field(name = "some-other-name")]
```

If you use uncased macro, then the payload HTML field name containing any case,
such as `HTML-FIELD-NAME`, will be matched to the struct field name.

For setting the default value, the syntax is as follows:

```
#[field(default = "default value")]
```

And, for validation, the syntax is as follows:

```
#[field(validate = validation_function())]
```

There are a couple of built-in validation functions in the `rocket::form::validate` module:

- `contains`: This function succeeds when the field as a string has this substring, or the field as a `Vec` contains this item, or `Option` has `Some(value)`, or `rocket::form::Result` has `Ok(value)` – for example, `contains("foo")`.

- `eq`: This function succeeds when the field value is equal to the function parameters. A type in Rust can be compared if the type implements `std::cmp::PartialEq`. You can see the example in the `NewUser` struct, `eq(self.password)`.

- `ext`: This function succeeds if the field type is `rocket::fs::TempFile` and the content type matches the function parameter – for example, `ext(rocket::http::ContentType::JavaScript)`.

- `len`: This function succeeds if the length of the field value is within the parameter range. You can see the example in our `NewUser` struct, `len(5..20)`. In Rust, we define the range as `from..to`, but we can omit the `to` part.

- `ne`: This function succeeds if the field value is not equal (`!=`) to the provided parameter. A type implementing the `std::cmp::PartialEq` trait can also use the inequality operator.

- `omits`: This function is the opposite of `contains`.

- `one_of`: This function succeeds if the value contains one of the items in the supplied parameter. The parameter must be an iterator.

- `range`: This function is like `len`, but it matches the value of the field instead of the value of the length of the field.

- `with`: We can pass a function or closure with the Boolean return type, and the function succeeds when the passed function or closure returns `true`.

Besides those functions, there are three more functions we can use. The functions work almost the same, but with different messages:

- `dbg_contains`: This function also returns the field value in the error message.

- `dbg_eq`: This function also returns the item value in the error message.

- `dbg_omits`: This function also returns the item value in the error message.

In the `NewUser` struct, we can see that we can also set a custom error message by combining the validation function with `.or_else("other message"`, as shown in the following example:

```
#[field(validate = len(5..20).or_else(msg!("name cannot
be empty")))]
```

Besides the provided functions, we can create a custom validation function. The function should return `form::Result<'_, ()>`. We want to implement custom validation for checking the strength of the password and the correctness of the email.

The first validation is password validation. We are going to use a crate called zxcvbn. This crate is a Rust port of the npm module of the same name created by Dropbox. The inspiration for the zxcvbn library is based on an **xkcd** comic strip (`https://xkcd.com/936/`), which said that random joined words we can remember, such as `"CorrectHorseBatteryStaple"`, are easier to remember and harder to crack compared to some rule such as *must contain a minimum of eight characters, of which one is upper case, one lower case, and one is a number.*

4. Add `zxcvbn = "2"` to the `Cargo.toml` dependencies, then create the following function in `src/models/user.rs`:

```rust
use rocket::form::{self, Error as FormError, FromForm};
use zxcvbn::zxcvbn;

...

fn validate_password(password: &str) -> form::Result<'_,
()> {
    let entropy = zxcvbn(password, &[]);
    if entropy.is_err() || entropy.unwrap().score()
    < 3 {
        return Err(FormError::validation("weak
        password").into());
    }
    Ok(())
}
```

You can set the scoring strength up to four, but it means we cannot send a weak password to the server. Right now, we just set the threshold of the password score to two.

5. After that, we can implement the validation for email correctness. First, add `regex` = `"1.5.4"` to `Cargo.toml` and add this function in `src/models/user.rs`:

```
use regex::Regex;

...

fn validate_email(email: &str) -> form::Result<'_, ()> {
    const EMAIL_REGEX: &str = r#"(?:[a-z0-9!#$%&
    '*+/=?^_`{|}~-]+(?:\.[a-z0-9!#$%&'*+/=?^_`{|}~-]
    +)*|"(?:[\x01-\x08\x0b\x0c\x0e-\x1f\x21\
    x23-\x5b\x5d-\x7f]|\\[\x01-\x09\x0b\x0c\
    x0e-\x7f])*")@(?:(?:[a-z0-9](?:[a-z0-9-]*[
    a-z0-9])?\.)+[a-z0-9](?:[a-z0-9-]*[a-
    z0-9])?|\[(?:(?:25[0-5]|2[0-4][0-9]|[01]?[0-9][
    0-9]?)\.){3}(?:25[0-5]|2[0-4][0-9]|[01]?[0-9][
    0-9]?|[a-z0-9-]*[a-z0-9]:(?:[\x01-\x08
    \x0b\x0c\x0e-\x1f\x21-\x5a\x53-\x7f]|\\[\x01
    -\x09\x0b\x0c\x0e-\x7f])+)\])"#;
    let email_regex = Regex::new(EMAIL_REGEX).
    unwrap();
    if !email_regex.is_match(email) {
        return Err(FormError::validation("invalid
        email").into());
    }
    Ok(())
}
```

6. If the email **regular expression (regex)** is hard to type, you can also copy and paste it from the source code on GitHub. We can now use the validations in the `NewUser` struct:

```
pub struct NewUser<'r> {
    ...
    #[field(validate = validate_email().
    or_else(msg!("invalid email")))]
    pub email: &'r str,
    #[field(validate = validate_password()
    .or_else(msg!("weak password")))]
    pub password: &'r str,
```

```
        . . .
    }
```

7. Next, we can implement the `create()` method for the `User` struct. For security purposes, we will hash the password using the secure password-hashing function. In 2021, people considered `md5` as a very insecure hashing function, and `sha1` and `sha3` as insecure hashing functions, so we will not use those functions. People usually consider using `bcrypt`, `scrypt`, or `argon2` instead. Now, `argon2` has a version, `argon2id`, which is resistant to side-channel attacks and GPU cracking attacks, so we will use `argon2` as a password-hashing implementation.

    There is another possible attack on the `create()` method: **cross-site scripting (XSS)**. For example, a user can input `"<script>console.log("hack")</script>"` as a description. We can rectify this problem by using an HTML sanitization library called `ammonia`.

    To add `argon2` and `ammonia` for the `create()` method, add these lines in `Cargo.toml`:

    ```
    ammonia = "3.1.2"
    argon2 = "0.3"
    rand_core = {version = "0.6", features = ["std"]}
    ```

8. We can create a function to sanitize HTML in `src/models/mod.rs`:

    ```
    use ammonia::Builder;
    use std::collections::hash_set::HashSet;


    pub fn clean_html(src: &str) -> String {
        Builder::default()
            .tags(HashSet::new())
            .clean(src)
            .to_string()
    }
    ```

    The default cleaner from `ammonia::Builder::default` allows many HTML tags, and people can still deface the site. To rectify this problem, we are passing an empty `HashSet` to disallow any HTML tag.

9. After the password hashing and HTML sanitization are ready, it is time to implement the `create()` method for the `User` struct. Add the required use directives in `src/models/user.rs`:

```
use super::clean_html;
use argon2::{password_hash::{rand_core::OsRng,
PasswordHasher, SaltString},Argon2};
```

10. Put these lines in the `impl User` block:

```
pub async fn create<'r>(
    connection: &mut PgConnection,
    new_user: &'r NewUser<'r>,
) -> Result<Self, Box<dyn Error>> {
    let uuid = Uuid::new_v4();
    let username = &(clean_html(new_user.username));
    let description = &(new_user.description.map(
    |desc| clean_html(desc)));
}
```

We generate a new UUID for the new `User` instance. After that, we clean the username value and description value. We don't clean the email and password because we already validate the content of the email using a regex and we will not show any password in the HTML.

11. Next, append the following lines to hash the password:

```
let salt = SaltString::generate(&mut OsRng);
let argon2 = Argon2::default();
let password_hash = argon2.hash_password(new_user.
password.as_bytes(), &salt);
if password_hash.is_err() {
    return Err("cannot create password hash".into());
}
```

12. Next, we send the `INSERT` statement to our database server and return the inserted row. Append the following lines:

```
let query_str = r#"INSERT INTO users
(uuid, username, email, password_hash, description,
status)
VALUES
```

```
($1, $2, $3, $4, $5, $6)
RETURNING *"#;
Ok(sqlx::query_as::<_, Self>(query_str)
    .bind(uuid)
    .bind(username)
    .bind(new_user.email)
    .bind(password_hash.unwrap().to_string())
    .bind(description)
    .bind(UserStatus::Inactive)
    .fetch_one(connection)
    .await?)
```

13. When the `User::create()` method is ready, we can implement the `create_user()` function. When the application successfully creates a user, it's better to show the result by redirecting to `get_user`. For this purpose, we can use the `rocket::response::Redirect` type instead of `RawHtml`.

    Also, if there's an error, it's better to redirect to `new_user()` and show the error so the user can fix the input error. We can do this by getting the error of the `NewUser` validations or any other error, and redirecting to the `new_user()` function with embedded error information.

    We can get the errors for the request form value using `rocket::form::Contextual`, a proxy for the form type that contains error information. We also going to use `rocket::response::Flash` to send a one-time cookie to the web browser and retrieve the message on a route using `rocket::request::FlashMessage`. Append these lines to `src/routes/user.rs`:

    ```
    use crate::models::{pagination::Pagination,
    user::{NewUser, User}};
    use rocket::form::{Contextual, Form};
    use rocket::request::FlashMessage;
    use rocket::response::{content::RawHtml, Flash,
    Redirect};
    ```

14. Change the signature of the `create_user()` function to the following:

    ```
    #[post("/users", format = "application/x-www-form-
    urlencoded", data = "<user_context>")]
    pub async fn create_user<'r>(
    ```

```
        mut db: Connection<DBConnection>,
        user_context: Form<Contextual<'r, NewUser<'r>>>,
    ) -> Result<Flash<Redirect>, Flash<Redirect>> {}
```

Because we are sending POST data, the browser will send Content-Type as
"application/x-www-form-urlencoded", so we have to change the
format accordingly.

Also, take a look at the request parameter; instead of Form<NewUser<'r>>,
we are inserting the Contextual type in the middle of the parameter.
We are also changing the return value to Result<Flash<Redirect>,
Flash<Redirect>>.

15. Now, let's implement the function body. Append the following lines to the
    function body:

```
if user_context.value.is_none() {
    let error_message = format!(
        "<div>{}</div>",
        user_context
            .context
            .errors()
            .map(|e| e.to_string())
            .collect::<Vec<_>>()
            .join("<br/>")
    );
    return Err(Flash::error(Redirect::to("/
    users/new"), error_message));
}
```

If user_context has value, it means that Rocket successfully converted
the request payload and put it inside the value attribute. We are branching
and returning Error with a Flash message and Redirect directive to "/
users/new".

16. The next implementation is if Rocket successfully parses NewUser. Append the
    following lines in the true branch:

```
let new_user = user_context.value.as_ref().unwrap();
let connection = db.acquire().await.map_err(|_| {
    Flash::error(
```

```
                    Redirect::to("/users/new"),
                    "<div>Something went wrong when creating
                    user</div>",
                )
        })?;
        let user = User::create(connection, new_user).await.map_
        err(|_| {
            Flash::error(
                Redirect::to("/users/new"),
                "<div>Something went wrong when creating
                user</div>",
            )
        })?;
        Ok(Flash::success(
            Redirect::to(format!("/users/{}", user.uuid)),
            "<div>Successfully created user</div>",
        ))
```

Just like the get_user() function, we create a routine to get the database
connection, perform an INSERT operation to the database server, and generate the
successful Redirect response. But, when an error occurs, instead of returning
HTML, we generate the Redirect directive with the appropriate path and message.

17. We now need to change the new_user() and get_user() functions to be able
    to handle incoming FlashMessage request guards. First, for the new_user()
    function, change the signature to the following:

```
get_user(
    mut db: Connection<DBConnection>,
    uuid: &str,
    flash: Option<FlashMessage<'_>>,
)
```

18. Because the flash message may not always exist, we wrap it in `Option`. After `let mut html_string = String::from(USER_HTML_PREFIX);`, append the following lines in the function body:

```
if flash.is_some() {
    html_string.push_str(flash.unwrap().message());
}
```

19. We change the `new_user()` function almost the same. Change the function signature to this:

```
new_user(flash: Option<FlashMessage<'_>>)
```

Then, append the following lines after `USER_HTML_PREFIX`:

```
if flash.is_some() {
    html_string.push_str(flash.unwrap().message());
}
```

20. Now, it's time to try creating the user data. If everything is correct, you should see screens like the following. The error message looks as follows:

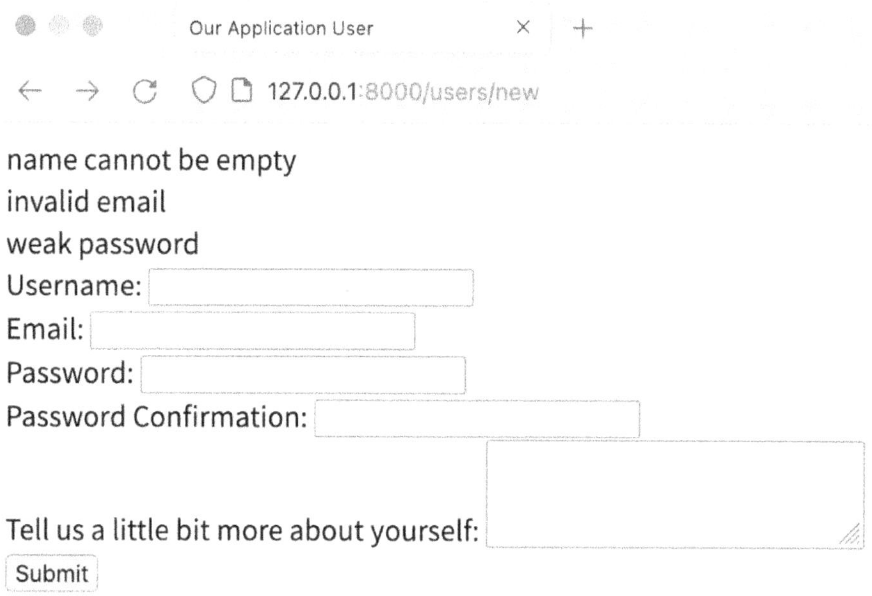

Figure 6.2 – Error message when failed

The success message looks like the following:

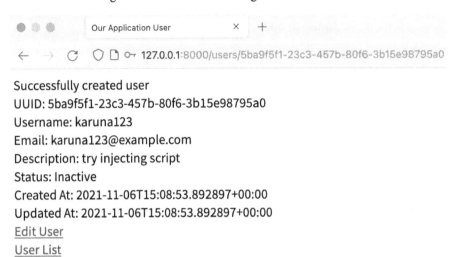

Successfully created user
UUID: 5ba9f5f1-23c3-457b-80f6-3b15e98795a0
Username: karuna123
Email: karuna123@example.com
Description: try injecting script
Status: Inactive
Created At: 2021-11-06T15:08:53.892897+00:00
Updated At: 2021-11-06T15:08:53.892897+00:00
Edit User
User List

Figure 6.3 – Success message

In the next sections, we are going to continue with updating the user and deleting a user.

Implementing PUT and PATCH user

To update the user, we need a page such as new_user(), but we want the form pre-populated with existing data. We also want to add another field for the user to confirm the old password. Let's look at the steps:

1. Change the edit_user() function signature to the following:

    ```
    #[get("/users/edit/<uuid>", format = "text/html")]
    pub async fn edit_user(mut db: Connection<DBConnection>,
    uuid: &str, flash: Option<FlashMessage<'_>>) ->
    HtmlResponse {}
    ```

2. To get the existing user, append the following lines inside the function body block:

    ```
    let connection = db
        .acquire()
        .await
        .map_err(|_| Status::InternalServerError)?;
    let user = User::find(connection, uuid)
        .await
        .map_err(|_| Status::NotFound)?;
    ```

3. After that, we can append the HTML, such as new_user(), but this time, we also include the existing data from the user. Append the following lines inside the edit_user() function body:

```
let mut html_string = String::from(USER_HTML_PREFIX);
if flash.is_some() {
    html_string.push_str(flash.unwrap().message());
}
html_string.push_str(
    format!(
        r#"<form accept-charset="UTF-8" action="/
        users/{}" autocomplete="off" method="POST">
<input type="hidden" name="_METHOD" value="PUT"/>
<div>
    <label for="username">Username:</label>
    <input name="username" type="text" value="{}"/>
</div>
<div>
    <label for="email">Email:</label>
    <input name="email" type="email" value="{}"/>
</div>
<div>
    <label for="old_password">Old password:</label>
    <input name="old_password" type="password"/>
</div>
<div>
    <label for="password">New password:</label>
    <input name="password" type="password"/>
</div>
<div>
    <label for="password_confirmation">Password
    Confirmation:</label>
    <input name="password_confirmation" type=
    "password"/>
</div>
<div>
    <label for="description">Tell us a little bit more
```

```
            about yourself:</label>
            <textarea name="description">{}</textarea>
        </div>
        <button type="submit" value="Submit">Submit</button>
        </form>"#,
                &user.uuid,
                &user.username,
                &user.email,
                &user.description.unwrap_or_else(||
                "".to_string()),
            )
            .as_ref(),
        );
        html_string.push_str(USER_HTML_SUFFIX);
        Ok(RawHtml(html_string))
```

After this, all previous links to `"/users/edit/{}"` that we implemented in the previous pages should work.

If you look at the code, we see the form's `method` attribute has a `"POST"` value. The reason is that the HTML standard says that a form method can only be GET or POST. Most web browsers will just change the invalid method, such as PUT and PATCH, to POST.

Some web frameworks work around this limitation by sending an **XMLHttpRequest** using JavaScript, or a newer API such as the JavaScript Fetch API. Other web applications send a POST request but include a hidden value in the request payload. We are going to use the second way to implement updating the user by adding a new field, name="_ METHOD", with the "PUT" value.

Just like `create_user()`, we want to execute `update_function()` to redirect to `edit_user()` again if there's something wrong. We also want to execute `update_ function()` to redirect to the user page again after successfully updating the user.

Since we are adding new payloads, _METHOD and old_password, we need a new type that is different from NewUser:

1. Create a new struct called EditedUser in src/models/user.rs:

    ```
    #[derive(Debug, FromForm)]
    pub struct EditedUser<'r> {
        #[field(name = "_METHOD")]
    ```

```
        pub method: &'r str,
        #[field(validate = len(5..20).or_else(msg!("name
        cannot be empty")))]
        pub username: &'r str,
        #[field(validate = validate_email()
        .or_else(msg!("invalid email")))]
        pub email: &'r str,
        pub old_password: &'r str,
        pub password: &'r str,
        pub password_confirmation: &'r str,
        #[field(default = "")]
        pub description: Option<&'r str>,
    }
```

2. We want to skip updating the password if there's no value in old_password,
    but if there is a value in old_password, we want to make sure the password
    strength is adequate and password_confirmation has the same content as
    password. Create a function in src/models/user.rs:

```
fn skip_validate_password<'v>(password: &'v str, old_
password: &'v str, password_confirmation: &'v str) ->
form::Result<'v, ()> {
    if old_password.is_empty() {
        return Ok(());
    }
    validate_password(password)?;
    if password.ne(password_confirmation) {
        return Err(FormError::validation("password
        confirmation mismatch").into());
    }
    Ok(())
}
```

3. Then, use the validation function in the directive above the password field:

```
#[field(validate = skip_validate_password(self.old_
password, self.password_confirmation))]
pub password: &'r str,
```

We need a method for `User` to update the database row based on the `EditedUser` content. This method will also verify the hash of `old_password` to make sure `EditedUser` is valid.

4. Add the use directive in `src/models/user.rs`:

```
use argon2::{password_hash::{rand_core::OsRng,
PasswordHash, PasswordHasher, PasswordVerifier,
SaltString}, Argon2};
use chrono::offset::Utc;
```

5. Create a new method inside the `impl User` block in `src/models/user.rs`:

```
pub async fn update<'r>(db: &mut Connection<DBConnection>,
uuid: &'r str, user: &'r EditedUser<'r>) -> Result<Self,
Box<dyn Error>> {}
```

6. Inside the method, fetch the old user data from the database:

```
let connection = db.acquire().await?;
let old_user = Self::find(connection, uuid).await?;
```

7. Prepare the data for updating:

```
let now = OurDateTime(Utc::now());
let username = &(clean_html(user.username));
let description = &(user.description.map(|desc| clean_
html(desc)));
```

8. Because we can change the password or skip changing the password depending on whether or not we have `old_password  or  not`, prepare the query items:

```
let mut set_strings = vec![
    "username = $1",
    "email = $2",
    "description = $3",
    "updated_at = $4",
];
```

```
let mut where_string = "$5";
let mut password_string = String::new();
let is_with_password = !user.old_password.is_empty();
```

9. If we are updating `password`, we need to verify `old_password` with the existing password. We also want to hash a new password and add the password to `set_strings`. Append the following lines:

```
if is_with_password {
    let old_password_hash = PasswordHash::
    new(&old_user.password_hash)
        .map_err(|_| "cannot read password hash")?;
    let argon2 = Argon2::default();
    argon2
        .verify_password(user.password.as_bytes(),
        &old_password_hash)
        .map_err(|_| "cannot confirm old password")?;
    let salt = SaltString::generate(&mut OsRng);
    let new_hash = argon2
        .hash_password(user.password.as_bytes(),
        &salt)
        .map_err(|_| "cannot create password hash")?;
    password_string.push_str(
    new_hash.to_string().as_ref());
    set_strings.push("password_hash = $5");
    where_string = "$6";
}
```

10. We can then construct the `UPDATE` statement to update the user, execute the statement, and return the `User` instance:

```
let query_str = format!(
    r#"UPDATE users SET {} WHERE uuid = {} RETURNING
    *"#,
    set_strings.join(", "),
    where_string,
);
let connection = db.acquire().await?;
```

```
    let mut binded = sqlx::query_as::<_, Self>(&query_str)
        .bind(username)
        .bind(user.email)
        .bind(description)
        .bind(&now);
if is_with_password {
    binded = binded.bind(&password_string);
}
    let parsed_uuid = Uuid::parse_str(uuid)?;
    Ok(binded.bind(parsed_uuid).fetch_one(connection).await?)
```

11. Now, it's time to use `EditedUser` and implement `update_user()`. Append `EditedUser` in the use directive:

```
use crate::models::{pagination::Pagination,
user::{EditedUser, NewUser, User}};
```

12. Create the `update_user()` function in `src/routes/user.rs`:

```
#[post("/users/<uuid>", format = "application/x-www-form-
urlencoded", data = "<user_context>")]
pub async fn update_user<'r>(db:
Connection<DBConnection>, uuid: &str, user_
context: Form<Contextual<'r, EditedUser<'r>>>) ->
Result<Flash<Redirect>, Flash<Redirect>> {}
```

13. In the function, we need to check whether the form is correct or not. Append the following lines:

```
if user_context.value.is_none() {
    let error_message = format!(
        "<div>{}</div>",
        user_context
            .context
            .errors()
            .map(|e| e.to_string())
            .collect::<Vec<_>>()
            .join("<br/>")
    );
    return Err(Flash::error(
```

```
        Redirect::to(format!("/users/edit/{}", uuid)),
        error_message,
    ));
}
```

14. We can tell the application to process depending on "_METHOD". Append the following lines:

```
let user_value = user_context.value.as_ref().unwrap();
match user_value.method {
    "PUT" => put_user(db, uuid, user_context).await,
    "PATCH" => patch_user(db, uuid, user_
    context).await,
    _ => Err(Flash::error(
        Redirect::to(format!("/users/edit/{}", uuid)),
        "<div>Something went wrong when updating
        user</div>",
    )),
}
```

We don't waste the functions we defined before. We are using the put_user() and patch_user() functions.

15. Now, it's time to implement the put_user() function. Change the signature of the put_user() function:

```
#[put("/users/<uuid>", format = "application/x-www-form-
urlencoded", data = "<user_context>")]
pub async fn put_user<'r>(mut db:
Connection<DBConnection>, uuid: &str, user_
context: Form<Contextual<'r, EditedUser<'r>>>) ->
Result<Flash<Redirect>, Flash<Redirect>> {}
```

Then, implement the function as follows:

```
let user_value = user_context.value.as_ref().unwrap();
let user = User::update(&mut db, uuid, user_value).await.
map_err(|_| {
    Flash::error(
        Redirect::to(format!("/users/edit/{}", uuid)),
        "<div>Something went wrong when updating
        user</div>",
```

```
        )
    })?;
    Ok(Flash::success(
        Redirect::to(format!("/users/{}", user.uuid)),
        "<div>Successfully updated user</div>",
    ))
```

16. For the `patch_user()` function, we can just reuse the `put_user()` function. Write the code for `patch_user()`:

```
#[patch("/users/<uuid>", format = "application/x-www-
form-urlencoded", data = "<user_context>")]
pub async fn patch_user<'r>(db: Connection<DBConnection>,
uuid: &str, user_context: Form<Contextual<'r,
EditedUser<'r>>>) -> Result<Flash<Redirect>,
Flash<Redirect>> {
    put_user(db, uuid, user_context).await
}
```

17. Finally, append the new route in `src/main.rs`:

```
user::edit_user,
user::update_user,
user::put_user,
```

The only endpoint left is for deleting the user. Let's continue with that in the next section.

Implementing DELETE user

The first thing we want to do to delete a user is to create a method for the `User` struct. Let's look at the steps:

1. Write the method to delete a user in the `impl User` block in `src/models/user.rs`:

```
pub async fn destroy(connection: &mut PgConnection, uuid:
&str) -> Result<(), Box<dyn Error>> {
    let parsed_uuid = Uuid::parse_str(uuid)?;
    let query_str = "DELETE FROM users WHERE uuid =
    $1";
    sqlx::query(query_str)
```

```
        .bind(parsed_uuid)
        .execute(connection)
        .await?;
    Ok(())
}
```

Then, we can implement the delete_user() function in src/routes/user.
rs:

```
#[delete("/users/<uuid>", format = "application/x-www-
form-urlencoded")]
pub async fn delete_user(
    mut db: Connection<DBConnection>,
    uuid: &str,
) -> Result<Flash<Redirect>, Flash<Redirect>> {
    let connection = db.acquire().await.map_err(|_| {
        Flash::error(
            Redirect::to("/users"),
            "<div>Something went wrong when deleting
            user</div>",
        )
    })?;
    User::destroy(connection, uuid).await.map_err(|_| {
        Flash::error(
            Redirect::to("/users"),
            "<div>Something went wrong when deleting
            user</div>",
        )
    })?;
    Ok(Flash::success(
        Redirect::to("/users"),
        "<div>Successfully deleted user</div>",
    ))
}
```

2. The problem is that neither the link nor the form in HTML are allowed to use the DELETE method. We cannot use the link, as any bot seeing it will crawl on it and could perform resource deletion accidentally. As with updating the user, we can use the form and send a POST request to a new endpoint. Add a new function in src/ routes/user.rs:

```
#[post("/users/delete/<uuid>", format = "application/x-
www-form-urlencoded")]
pub async fn delete_user_entry_point(
    db: Connection<DBConnection>,
    uuid: &str,
) -> Result<Flash<Redirect>, Flash<Redirect>> {
    delete_user(db, uuid).await
}
```

3. Don't forget to add the route in src/main.rs:

```
user::delete_user,
user::delete_user_entry_point,
```

4. Now, where can we create a form to delete the user? Let's do that on the get_ user() page. Append the HTML for the form as follows:

```
html_string
    .push_str(format!(r#"<a href="/users/edit/{}">Edit
    User</a><br/>"#, user.uuid).as_ref());
html_string.push_str(
    format!(
        r#"<form accept-charset="UTF-8" action="/
        users/delete/{}" autocomplete="off"
        method="POST"><button type="submit"
        value="Submit">Delete</button></form>"#,
        user.uuid
    )
    .as_ref(),
);
```

We have now completed all of the endpoints used to manage users. Try adding users and see how pagination works or try improving the HTML. You can also try activating users for a challenge!

Summary

In this chapter, we have learned about the basic operations for user entities by implementing the creation, reading, updating, and deleting of user routes.

We also learned more about various modules of the Rocket framework such as RawHtml, Redirect, Contextual, Flash, Form, and FlashMessage.

Along with implementing the endpoints, we also learned more about database operations such as querying, inserting, updating, and deleting objects on a database server.

In the next chapter, we are going to learn more about error handling and creating our own error types.

7
Handling Errors in Rust and Rocket

In the previous chapter, we learned about creating endpoints and SQL queries to handle the management of the User entity. In this chapter, we are going to learn more about error handling in Rust and Rocket. After learning the concepts in this chapter, you will be able to implement error handling in a Rocket application.

We are also going to discuss more common ways to handle errors in Rust and Rocket, including signaling unrecoverable errors using the panic! macro and catching the panic! macro, using Option, using Result, creating a custom Error type, and logging the generated error.

In this chapter, we're going to cover the following main topics:

- Using panic!
- Using Option
- Returning Result
- Creating a custom error type
- Logging errors

Technical requirements

For this chapter, we have the same technical requirements as the previous chapter. We need a Rust compiler, a text editor, an HTTP client, and a PostgreSQL database server.

You can find the source code for this chapter at `https://github.com/PacktPublishing/Rust-Web-Development-with-Rocket/tree/main/Chapter07`.

Using panic!

To understand error handling in Rust, we need to begin with the `panic!` macro. We can use the `panic!` macro when the application encounters an unrecoverable error and there's no purpose in continuing the application. If the application encounters `panic!`, the application will emit the backtrace and terminate.

Let's try using `panic!` on the program that we created in the previous chapter. Suppose we want the application to read a secret file before we initialize Rocket. If the application cannot find this secret file, it will not continue.

Let's get started:

1. Add the following line in `src/main.rs`:

    ```
    use std::env;
    ```

2. In the same file in the `rocket()` function, prepend the following lines:

    ```
    let secret_file_path = env::current_dir().unwrap().
    join("secret_file");
    if !secret_file_path.exists() {
        panic!("secret does not exists");
    }
    ```

3. Afterward, try executing `cargo run` without creating an empty file named `secret_file` inside the working directory. You should see the output as follows:

    ```
    thread 'main' panicked at 'secret does not exists', src/
    main.rs:15:9
    note: run with `RUST_BACKTRACE=1` environment variable to
    display a backtrace
    ```

4. Now, try running the application again with RUST_BACKTRACE=1 cargo run. You should see the backtrace output in the terminal similar to the following:

```
RUST_BACKTRACE=1 cargo run
    Finished dev [unoptimized + debuginfo] target(s)
    in 0.18s
     Running `target/debug/our_application`
thread 'main' panicked at 'secret does not exists', src/
main.rs:15:9
stack backtrace:
...
  14: our_application::main
            at ./src/main.rs:12:36
  15: core::ops::function::FnOnce::call_once
            at /rustc/59eed8a2aac0230a8b5
            3e89d4e99d55912ba6b35/library/core/
            src/ops/function.rs:227:5
note: Some details are omitted, run with `RUST_
BACKTRACE=full` for a verbose backtrace.
```

5. Sometimes, we don't want to deallocate after panicking using the panic! macro because we want the application to exit as soon as possible. We can skip deallocating by setting panic = "abort" in Cargo.toml under the profile we are using. Setting that configuration will make our binary smaller and exit faster, and the operating system will need to clean it later. Let's try doing that. Set the following lines in Cargo.toml and run the application again:

```
[profile.dev]
panic = "abort"
```

Now that we know how to use panic!, let's see how we can catch it in the next section.

Catching panic!

As well as using panic!, we can also use the todo! and unimplemented! macros in Rust code. Those macros are very useful for prototyping because they will call panic! while also allowing the code to type-check at compile time.

But, why does Rocket not shut down when we are calling a route with `todo!`? If we check the Rocket source code, there's a `catch_unwind` function in `src::panic` that can be used to capture a panicking function. Let's see that code in the Rocket source code, `core/lib/src/server.rs`:

```
let fut = std::panic::catch_unwind(move || run())
          .map_err(|e| panic_info!(name, e))
          .ok()?;
```

Here, `run()` is a route handling function. Each time we call a route that is panicking, the preceding routine will convert the panic into the result's `Err` variant. Try removing the `secret_file_path` routine we added before and running the application. Now, create a user and try going into user posts. For example, create a user with the `95a54c16-e830-45c9-ba1d-5242c0e4c18f` UUID. Try opening `http://127.0.0.1/users/95a54c16-e830-45c9-ba1d-5242c0e4c18f/posts`. Since we only put `todo!("will implement later")` in the function body, the application will panic, but the preceding `catch_unwind` function will catch the panic and convert it into an error. Please note that `catch_unwind` will not work if we set `panic = "abort"` in `Cargo.toml`.

In a regular workflow, we don't usually want to use `panic!`, because panicking interrupts everything, and the program will not be able to continue. If the Rocket framework does not catch `panic!` and one of the route handling functions is panicking, then that single error will close the application and there will be nothing to handle the other requests. But, what if we want to terminate the Rocket application when we encounter an unrecoverable error? Let's see how we can do it in the next section.

Using shutdown

To shut down smoothly if the application encounters an unrecoverable error in the route handling function, we can use the `rocket::Shutdown` request guard. Remember, the request guard is a parameter we are supplying to the route handling functions.

To see the `Shutdown` request guard in action, let's try implementing it in our application. Using the previous application, add a new route in `src/routes/mod.rs` called `/shutdown`:

```
use rocket::Shutdown;

...

#[get("/shutdown")]
pub async fn shutdown(shutdown: Shutdown) -> &'static str {
    // suppose this variable is from function which
```

```
    // produces irrecoverable error
    let result: Result<&str, &str> = Err("err");
    if result.is_err() {
        shutdown.notify();
        return "Shutting down the application.";
    }
    return "Not doing anything.";
}
```

Try adding the shutdown() function in src/main.rs. After that, rerun the application and send an HTTP request to /shutdown while monitoring the output of the application on the terminal. The application should shut down smoothly.

In the next two sections, let's see how we can use Option and Result as an alternative way to handle errors.

Using Option

In programming, a routine might produce a correct result or encounter a problem. One classical example is division by zero. Dividing something by zero is mathematically undefined. If the application has a routine to divide something, and the routine encounters zero as input, the application cannot return any number. We want the application to return another type instead of a number. We need a type that can hold multiple variants of data.

In Rust, we can define an enum type, a type that can be different variants of data. An enum type might be as follows:

```
enum Shapes {
    None,
    Point(i8),
    Line(i8, i8),
    Rectangle {
        top: (i8, i8),
        length: u8,
        height: u8,
    },
}
```

Point and Line are said to have **unnamed fields**, while Rectangle is said to have **named fields**. Rectangle can also be called a **struct-like enum** variant.

If all members of enum have no data, we can add a discriminant on the member. Here is an example:

```
enum Color {
    Red,            // 0
    Green = 127,  // 127
    Blue,           // 128
}
```

We can assign enum to a variable, and use the variable in a function as in the following:

```
fn do_something(color: Color) -> Shapes {
    let rectangle = Shapes::Rectangle {
        top: (0, 2),
        length: 10,
        height: 8,
    };
    match color {
        Color::Red => Shapes::None,
        Color::Green => Shapes::Point(10),
        _ => rectangle,
    }
}
```

Going back to error handling, we can use enum to communicate that there's something wrong in our code. Going back to division by zero, here is an example:

```
enum Maybe {
    WeCannotDoIt,
    WeCanDoIt(i8),
}


fn check_divisible(input: i8) -> Maybe {
    if input == 0 {
        return Maybe::WeCannotDoIt;
    }
```

```
    Maybe::WeCanDoIt(input)
}
```

The preceding pattern returning something or not returning something is very common, so Rust has its own enum to show whether we have something or not in the standard library, called `std::option::Option`:

```
pub enum Option<T> {
    None,
    Some(T),
}
```

`Some(T)` is used to communicate that we have `T`, and `None` is obviously used to communicate that we don't have `T`. We used `Option` in some of the previous code. For example, we used it in the `User` struct:

```
struct User {
    ...
    description: Option<String>,
    ...
}
```

We also used `Option` as a function parameter or return type:

```
find_all(..., pagination: Option<Pagination>) -> (...,
Option<Pagination>), ... {}
```

There are many useful things we can use with `Option`. Suppose we have two variables, `we_have_it` and `we_do_not_have_it`:

```
let we_have_it: Option<usize> = Some(1);
let we_do_not_have_it: Option<usize> = None;
```

- One thing we can do is pattern matching and use the content:

```
    match we_have_it {
        Some(t) => println!("The value = {}", t),
        None => println!("We don't have it"),
    };
```

- We can process it in a more convenient way if we care about the content of we_ have_it:

```
if let Some(t) = we_have_it {
    println!("The value = {}", t);
}
```

- Option can be compared if the inner type implements std::cmp::Eq and std::cmp::Ord, that is, the inner type can be compared using ==, !=, >, and other comparison operators. Notice that we use assert!, a macro used for testing:

```
assert!(we_have_it != we_do_not_have_it);
```

- We can check whether a variable is Some or None:

```
assert!(we_have_it.is_some());
assert!(we_do_not_have_it.is_none());
```

- We can also get the content by unwrapping Option. But, there's a caveat; unwrapping None will cause panic, so be careful when unwrapping Option. Notice we use assert_eq!, which is a macro used for testing to ensure equality:

```
assert_eq!(we_have_it.unwrap(), 1);
// assert_eq!(we_do_not_have_it.unwrap(), 1);
// will panic
```

- We can also use the expect() method. This method will work the same with unwrap() but we can use a custom message:

```
assert_eq!(we_have_it.expect("Oh no!"), 1);
// assert_eq!(we_do_not_have_it.expect("Oh no!"), 1); //
will panic
```

- We can unwrap and set the default value so it will not panic if we unwrap None:

```
assert_eq!(we_have_it.unwrap_or(42), 1);
assert_eq!(we_do_not_have_it.unwrap_or(42), 42);
```

- We can unwrap and set the default value with a closure:

```
let x = 42;
assert_eq!(we_have_it.unwrap_or_else(|| x), 1);
assert_eq!(we_do_not_have_it.unwrap_or_else(|| x), 42);
```

- We can convert the value contained to something else using `map()`, `map_or()`, or `map_or_else()`:

```
assert_eq!(we_have_it.map(|v| format!("The value = {}",
v)), Some("The value = 1".to_string()));
assert_eq!(we_do_not_have_it.map(|v| format!("The value =
{}", v)), None);
assert_eq!(we_have_it.map_or("Oh no!".to_string(),
|v| format!("The value = {}", v)), "The value = 1".to_
string());
assert_eq!(we_do_not_have_it.map_or("Oh no!".to_string(),
|v| format!("The value = {}", v)), "Oh no!".to_string());
assert_eq!(we_have_it.map_or_else(|| "Oh no!".to_
string(), |v| format!("The value = {}", v)), "The value =
1".to_string());
assert_eq!(we_do_not_have_it.map_or_else(|| "Oh no!".
to_string(), |v| format!("The value = {}", v)), "Oh no!".
to_string());
```

There are other important methods, which you can check in the documentation for `std::option::Option`. Even though we can use `Option` to handle a situation where there's something or nothing, it does not convey a message of *something went wrong*. We can use another type similar to `Option` in the next part to achieve this.

Returning Result

In Rust, we have the `std::result::Result` enum that works like `Option`, but instead of saying *we have it* or *we don't have it*, the `Result` type is more about saying *we have it* or *we have this error*. Just like `Option`, `Result` is an enum type of the possible T type or possible E error:

```
enum Result<T, E> {
    Ok(T),
    Err(E),
}
```

Going back to the division by zero problem, take a look at the following simple example:

```
fn division(a: usize, b: usize) -> Result<f64, String> {
    if b == 0 {
        return Err(String::from("division by zero"));
    }
    return Ok(a as f64 / b as f64);
}
```

We don't want division by 0, so we return an error for the preceding function.

Similar to Option, Result has many convenient features we can use. Suppose we have the we_have_it and we_have_error variables:

```
let we_have_it: Result<usize, &'static str> = Ok(1);
let we_have_error: Result<usize, &'static str> = Err("Oh no!");
```

- We can get the value or the error using pattern matching:

```
match we_have_it {
    Ok(v) => println!("The value = {}", v),
    Err(e) => println!("The error = {}", e),
};
```

- Or, we can use if let to destructure and get the value or the error:

```
if let Ok(v) = we_have_it {
    println!("The value = {}", v);
}
if let Err(e) = we_have_error {
    println!("The error = {}", e);
}
```

- We can compare the Ok variant and the Err variant:

```
assert!(we_have_it != we_have_error);
```

- We can check whether a variable is an Ok variant or an Err variant:

```
assert!(we_have_it.is_ok());
assert!(we_have_error.is_err());
```

- We can convert Result to Option:

```
assert_eq!(we_have_it.ok(), Some(1));
assert_eq!(we_have_error.ok(), None);
assert_eq!(we_have_it.err(), None);
assert_eq!(we_have_error.err(), Some("Oh no!"));
```

- Just like Option, we can use unwrap(), unwrap_or(), or unwrap_or_else():

```
assert_eq!(we_have_it.unwrap(), 1);
// assert_eq!(we_have_error.unwrap(), 1);
// panic
assert_eq!(we_have_it.expect("Oh no!"), 1);
// assert_eq!(we_have_error.expect("Oh no!"), 1);
// panic
assert_eq!(we_have_it.unwrap_or(0), 1);
assert_eq!(we_have_error.unwrap_or(0), 0);
assert_eq!(we_have_it.unwrap_or_else(|_| 0), 1);
assert_eq!(we_have_error.unwrap_or_else(|_| 0), 0);
```

- And, we can use map(), map_err(), map_or(), or map_or_else():

```
assert_eq!(we_have_it.map(|v| format!("The value = {}",
v)), Ok("The value = 1".to_string()));
assert_eq!(
    we_have_error.map(|v| format!("The error = {}",
    v)),
    Err("Oh no!")
);
assert_eq!(we_have_it.map_err(|s| s.len()), Ok(1));
assert_eq!(we_have_error.map_err(|s| s.len()), Err(6));
assert_eq!(we_have_it.map_or("Default value".to_string(),
|v| format!("The value = {}", v)), "The value = 1".to_
string());
assert_eq!(we_have_error.map_or("Default value".to_
string(), |v| format!("The value = {}", v)), "Default
value".to_string());
assert_eq!(we_have_it.map_or_else(|_| "Default value".to_
string(), |v| format!("The value = {}", v)), "The value =
```

```
1".to_string());
assert_eq!(we_have_error.map_or_else(|_| "Default value".
to_string(), |v| format!("The value = {}", v)), "Default
value".to_string());
```

There are other important methods besides those methods in the
std::result::Result documentation. Do check them because Option and
Result are very important in Rust and Rocket.

Returning a string or numbers as an error might be acceptable in some cases, but most
likely, we want a real error type with a message and possible backtrace that we can process
further. In the next section, we are going to learn about (and use) the Error trait and
return the dynamic error type in our application.

Creating a custom error type

Rust has a trait to unify propagating errors by providing the std::error::Error trait.
Since the Error trait is defined as pub trait Error: Debug + Display, any
type that implements Error should also implement the Debug and Display traits.

Let's see how we can create a custom error type by creating a new module:

1. In src/lib.rs, add the new errors module:

    ```
    pub mod errors;
    ```

2. After that, create a new folder, src/errors, and add the src/errors/mod.
    rs and src/errors/our_error.rs files. In src/errors/mod.rs, add this
    line:

    ```
    pub mod our_error;
    ```

3. In src/errors/our_error.rs, add the custom type for error:

    ```
    use rocket::http::Status;
    use std::error::Error;
    use std::fmt;

    #[derive(Debug)]
    pub struct OurError {
        pub status: Status,
        pub message: String,
        debug: Option<Box<dyn Error>>,
    ```

```
    }

    impl fmt::Display for OurError {
        fn fmt(&self, f: &mut fmt::Formatter<'_>) ->
        fmt::Result {
            write!(f, "{}", &self.message)
        }
    }
```

4. Then, we can implement the `Error` trait for `OurError`. In `src/errors/our_error.rs`, add the following lines:

```
    impl Error for OurError {
        fn source(&self) -> Option<&(dyn Error + 'static)> {
            if self.debug.is_some() {
                self.debug.as_ref().unwrap().source();
            }
            None
        }
    }
```

Currently, for the `User` module, we return a `Result<..., Box<dyn Error>>` dynamic error for each method. This is a common pattern of returning an error by using any type that implements `Error` and then putting the instance in the heap using `Box`.

The problem with this approach is we can only use methods provided by the `Error` trait, that is, `source()`. We want to be able to use the `OurError` status, message, and debug information.

5. So, let's add a couple of builder methods to `OurError`. In `src/errors/our_error.rs`, add the following lines:

```
    impl OurError {
        fn new_error_with_status(status: Status, message:
        String, debug: Option<Box<dyn Error>>) -> Self {
            OurError {
                status,
                message,
                debug,
```

```
        }
    }

    pub fn new_bad_request_error(message: String,
    debug: Option<Box<dyn Error>>) -> Self {
        Self::new_error_with_status(Status::
        BadRequest, message, debug)
    }

    pub fn new_not_found_error(message: String,
    debug: Option<Box<dyn Error>>) -> Self {
        Self::new_error_with_status(Status::NotFound,
        message, debug)
    }

    pub fn new_internal_server_error(
        message: String,
        debug: Option<Box<dyn Error>>,
    ) -> Self {
        Self::new_error_with_status(Status::
        InternalServerError, message, debug)
    }
}
```

6. If we take a look at `src/models/user.rs`, there are three sources of errors: `sqlx::Error`, `uuid::Error`, and argon2. Let's create a conversion for `sqlx::Error` and `uuid::Error` to `OurError`. Add the following use directive in `src/errors/our_error.rs`:

```
use sqlx::Error as sqlxError;
use uuid::Error as uuidError;
```

7. Inside the same file, `src/errors/our_error.rs`, add the following lines:

```
impl OurError {

    ...

    pub fn from_uuid_error(e: uuidError) -> Self {
        OurError::new_bad_request_error(
```

```
            String::from("Something went wrong"),
            Some(Box::new(e)))
    }
}
```

8. For `sqlx::Error`, we want to convert `not_found` error to HTTP status `404`
    and duplicate index error to an HTTP status 400bad request. Add the following
    lines to `src/errors/our_error.rs`:

```rust
use std::borrow::Cow;

....

impl OurError {

    ....

    pub fn from_sqlx_error(e: sqlxError) -> Self {
        match e {
            sqlxError::RowNotFound => {
                OurError::new_not_found_error(
                    String::from("Not found"),
                    Some(Box::new(e)))
            }
            sqlxError::Database(db) => {
                if db.code().unwrap_or(Cow::
                Borrowed("2300")).starts_with("23") {
                    return OurError::new_bad_
                    request_error(
                        String::from("Cannot create or
                        update resource"),
                        Some(Box::new(db)),
                    );
                }
                OurError::new_internal_server_error(
                    String::from("Something went
                    wrong"),
                    Some(Box::new(db)),
                )
            }
            _ => OurError::new_internal_server_error(
```

```
                    String::from("Something went wrong"),
                    Some(Box::new(e)),
            ),
        }
    }
}
```

9. We need to do one more thing before we modify our `User` entity. Some crates in Rust do not compile the `std` library by default to make the resulting binary smaller and embeddable in IoT (Internet of Things) devices or WebAssembly. For example, the `argon2` crate does not include the `Error` trait implementation by default, so we need to enable the `std` feature. In `Cargo.toml`, modify the `argon2` dependencies to enable the `std` library features:

    ```
    argon2 = {version = "0.3", features = ["std"]}
    ```

10. In `src/models/user.rs`, delete `use std::error::Error;` and replace it with `use crate::errors::our_error::OurError;`. Then, we can replace the methods for `User` to use `OurError` instead. Here is an example:

    ```
    pub async fn find(connection: &mut PgConnection, uuid:
    &str) -> Result<Self, OurError> {
        let parsed_uuid = Uuid::parse_str(
        uuid).map_err(OurError::from_uuid_error)?;
        let query_str = "SELECT * FROM users WHERE uuid =
        $1";
        Ok(sqlx::query_as::<_, Self>(query_str)
            .bind(parsed_uuid)
            .fetch_one(connection)
            .await
            .map_err(OurError::from_sqlx_error)?)
    }
    ```

11. For the `argon2` error, we can create a function or method, or convert it manually. For example, in `src/models/user.rs`, we can do this:

    ```
    let password_hash = argon2
        .hash_password(new_user.password.as_bytes(),
        &salt)
        .map_err(|e| {
    ```

```
OurError::new_internal_server_error(
    String::from("Something went wrong"),
    Some(Box::new(e)),
)
})?;
```

Change all the methods to use OurError. Just a reminder: you can find the complete source code for src/models/user.rs in the GitHub repository at https://github.com/PacktPublishing/Rust-Web-Development-with-Rocket/tree/main/Chapter07.

12. We will then use the OurError status and message in src/routes/user.rs. Because the Error type already implements the Display trait, we can use e directly inside format!(). Here is an example:

```
pub async fn get_user(...) -> HtmlResponse {
...
    let user = User::find(connection,
    uuid).await.map_err(|e| e.status)?;
...
}
...
pub async fn delete_user(...) -> Result<Flash<Redirect>,
Flash<Redirect>> {
...
    User::destroy(connection, uuid)
        .await
        .map_err(|e| Flash::error(Redirect::to("/
        users"), format!("<div>{}</div>", e)))?;
...
}
```

You can find the complete source code for src/routes/user.rs in the GitHub repository. Now that we have implemented errors, it might be a good time to try to implement the catchers that we defined before in src/catchers/mod.rs to show default errors for the user. You can also see the example of the default catchers in the source code.

In an application, tracking and logging errors are an important part of maintaining the application. Since we implemented the `Error` trait, we can log the `source()` of an error in our application. Let's take a look at how to do that in the next section.

Logging errors

In Rust, there's a log crate that provides a facade for application logging. The log provides five macros: `error!`, `warn!`, `info!`, `debug!`, and `trace!`. An application can then create a log based on the severity and filter what needs to be logged, also based on the severity. For example, if we filter based on `warn`, then we only log `error!` and `warn!` and ignore the rest. Since the log crate does not implement the logging itself, people often use another crate to do the actual implementation. In the documentation for the log crate, we can find examples of other logging crates that can be used: `env_logger`, `simple_logger`, `simplelog`, `pretty_env_logger`, `stderrlog`, `flexi_logger`, `log4rs`, `fern`, `syslog`, and `slog-stdlog`.

Let's implement custom logging in our application. We will use the `fern` crate for logging and wrap that in `async_log` to make logging asynchronous:

1. First, add these crates in `Cargo.toml`:

    ```
    async-log = "2.0.0"
    fern = "0.6"
    log = "0.4"
    ```

2. In `Rocket.toml`, add the config for `log_level`:

    ```
    log_level = "normal"
    ```

3. We can then create the function to initialize a global logger in our application. In `src/main.rs`, create a new function called `setup_logger`:

    ```
    fn setup_logger() {}
    ```

4. Inside the function, let's initialize the logger:

```
use log::LevelFilter;

...

let (level, logger) = fern::Dispatch::new()
    .format(move |out, message, record| {
        out.finish(format_args!(
            "[{date}] [{level}] [{target}] [{
             message}]",
            date = chrono::Local::now().format("[
            %Y-%m-%d] [%H:%M:%S%.3f]"),
            target = record.target(),
            level = record.level(),
            message = message
        ))
    })
    .level(LevelFilter::Info)
    .chain(std::io::stdout())
    .chain(
        fern::log_file("logs/application.log")
            .unwrap_or_else(|_| panic!("Cannot open
            logs/application.log")),
    )
    .into_log();
```

First, we create a new instance of fern::Dispatch. After that, we configure the output format using the format() method. After setting the output format, we set the log level using the level() method.

For the logger, we want to not only output the log to the operating system stdout, but we also want to write to a log file. We can do it using the chain() method. To avoid panicking, don't forget to create a logs folder in the application directory.

5. After we set up the level and logger, we wrap it inside async_log:

```
async_log::Logger::wrap(logger, || 0).start(level).
unwrap();
```

6. We will log `OurError` when it's created. Inside `src/errors/our_error.rs`, add the following lines:

```
impl OurError {
    fn new_error_with_status(...) ... {
        if debug.is_some() {
            log::error!("Error: {:?}", &debug);
        }
        ...
    }
}
```

7. Add the `setup_logger()` function to `src/main.rs`:

```
async fn rocket() -> Rocket<Build> {
    setup_logger();
    ...
}
```

8. Now, let's try to see `OurError` inside the application log. Try creating users with the same username; the application should emit a duplicate username error in the terminal and `logs/application.log` similar to the following:

```
[[2021-11-21][17:50:49.366]] [ERROR][our_
application::errors::our_error]
[Error: Some(PgDatabaseError { severity: Error, code:
"23505", message:
"duplicate key value violates unique constraint \"users_
username_key\""
, detail: Some("Key (username)=(karuna) already
exists."), hint: None, p
osition: None, where: None, schema: Some("public"),
table: Some("users")
, column: None, data_type: None, constraint: Some("users_
username_key"),
file: Some("nbtinsert.c"), line: Some(649), routine:
Some("_bt_check_un
ique") })]
```

Now that we have learned how to log errors, we can implement logging functionalities to improve the application. For example, we might want to create server-side analytics, or we can combine the logs with third-party monitoring as a service to improve the operations and create business intelligence.

Summary

In this chapter, we have learned some ways to handle errors in Rust and Rocket applications. We can use `panic!`, `Option`, and `Result` as a way to propagate errors and create handling for the errors.

We have also learned about creating a custom type that implements the `Error` trait. The type can store another error, creating an error chain.

Finally, we learned ways to log errors in our application. We can also use log capability to improve the application itself.

Our user pages are looking good, but using `String` all over the place is cumbersome, so in the next chapter, we are going to learn more about templating using CSS, JavaScript, and other assets in our application.

8
Serving Static Assets and Templates

One of the common functions of a web application is serving static files such as **Cascading Style Sheets** (**CSS**) or **JavaScript** (**JS**) files. In this chapter, we are going to learn about serving static assets from the Rocket application.

One common task for a web framework is rendering a template into HTML files. We are going to learn about using the Tera template to render HTML from the Rocket application.

In this chapter, we're going to cover the following main topics:

- Serving static assets
- Introducing the Tera template
- Showcasing users
- Working with forms
- Securing HTML forms from CSRF

Technical requirements

For this chapter, we have the same technical requirements as the previous chapter. We need a Rust compiler, a text editor, an HTTP client, and a PostgreSQL database server.

For the text editor, you can try adding an extension supporting the Tera template. If there is no extension for Tera, try adding an extension for a Jinja2 or Django template and set the file association to include the `*.tera` file.

We are going to add CSS to our application, and we are going to use stylesheets from `https://minicss.org/` since it's small and open source. Feel free to use and modify the example HTML with other stylesheets.

You can find the source code for this chapter at `https://github.com/PacktPublishing/Rust-Web-Development-with-Rocket/tree/main/Chapter08`.

Serving static assets

Serving static assets (such as HTML files, JS files, or CSS files) is a very common task for a web application. We can make Rocket serve files as well. Let's create the first function to serve a favicon. Previously you might have noticed that some web browsers requested a favicon file from our server, even though we did not explicitly mention it on our served HTML page. Let's look at the steps:

1. In the application root folder, create a folder named `static`. Inside the `static` folder, add a file named `favicon.png`. You can find sample `favicon.png` files on the internet or use the file from the sample source code for this chapter.

2. In `src/main.rs`, add a new route:

    ```
    routes![
        ...
        routes::favicon,
    ],
    ```

3. In `src/routes/mod.rs`, add a new route handling function to serve `favicon.png`:

    ```
    use rocket::fs::{relative, NamedFile};
    use std::path::Path;
    ...
    #[get("/favicon.png")]
    pub async fn favicon() -> NamedFile {
    ```

```
NamedFile::open(Path::new(relative!("static/
favicon.png")))
    .await
    .ok()
    .unwrap()
}
```

Here, `relative!` is a macro that generates a crate-relative version of a path. This means that the macro refers to the folder of the source file or the generated binary. For example, we have the source file for this application in `/some/source`, and by saying `relative!("static/favicon.png")`, it means the path is `/some/source/static/favicon.png`.

Every time we want to serve a particular file, we can create a route handling function, return `NamedFile`, and mount the route to Rocket. But obviously, this approach is not good; we can create a function to return static files dynamically.

4. Let's reuse the `assets` function that we created when we made the application skeleton. Previously, in *Chapter 3*, *Rocket Requests and Responses*, we learned that we can use multiple segments in a route. We can leverage this and serve a file that has the same filename with the request's multiple segments.

    Delete the favicon function that we created earlier and remove the reference to the function from `src/main.rs`. In `src/routes/mod.rs`, modify the use declarations:

    ```
    use rocket::fs::{relative, NamedFile};
    use std::path::{Path, PathBuf};
    ```

5. The application should return an HTTP `404` status code if the application cannot find the requested file. We can easily return `404` status code by wrapping `NamedFile` inside `Option`. If `NamedFile` is `None`, then the response will have `404` status automatically. Modify the `assets` function signature in `src/routes/mod.rs`:

    ```
    #[get("/<filename..>")]
    pub async fn assets(filename: PathBuf) ->
    Option<NamedFile> {}
    ```

6. We can then implement the `assets` function:

    ```
    let mut filename = Path::new(relative!("static")).
    join(filename);
    NamedFile::open(filename).await.ok()
    ```

7. Unfortunately, Rocket returns an HTTP 200 status code if filename is a directory, so an attacker can try attacking and mapping the folder structure inside the static folder. Let's handle this case by adding these lines:

```
let mut filename = Path::new(relative!("static")).
join(filename);
if filename.is_dir() {
    filename.push("index.html");
}
NamedFile::open(filename).await.ok()
```

If an attacker tries to systematically check the path inside the static file, the attacker will be served with an HTTP 404 status code and will not be able to infer the folder structures inside the static folder.

8. There's another way to serve the static file: by using the built-in rocket::fs::FileServer struct. Remove the function to handle static assets in src/routes/mod.rs, and append the following lines in src/main.rs:

```
use rocket::fs::relative;
use rocket::fs::FileServer;

...

.mount("/assets", FileServer::from(
  relative!("static")))
```

Even though web frameworks such as Rocket can serve static files, it's more common to serve static files behind web servers such as Apache Web Server or NGINX. In more advanced setups, people also utilize cloud storage, such as Amazon Web Services S3 or Google Cloud Storage, in conjunction with a **Content Delivery Network** (**CDN**).

In the next section, we are going to refine the HTML that we created in *Chapter 6, Implementing User CRUD*.

Introducing the Tera template

In web applications, there's usually a part that works as a web template system. Web designers and web developers can create web templates, and the web application generates HTML pages from the templates.

There are different kinds of web templates: server-side web templates (in which the template is rendered on the server-side), client-side web templates (where client-side applications render the template), or hybrid web templates.

There are a couple of templating engines in Rust. We can find templating engines for web development (such as **Handlebars, Tera, Askama**, or **Liquid**) at `https://crates.io` or `https:/lib.rs`.

The Rocket web framework has built-in support for templating in the form of the `rocket_dyn_templates` crate. Currently, the crate only supports two engines: **Handlebars** and **Tera**. In this book, we are going to use Tera as the template engine to simplify the development, but feel free to try the Handlebars engine as well.

Tera is a template engine that is inspired by **Jinja2** and **Django** templates. You can find the documentation for Tera at `https://tera.netlify.app/docs/`. A Tera template is a text file with expressions, statements, and comments. The expressions, statements, and comments are replaced with variables and expressions when the template is rendered.

For example, let's say we have a file named `hello.txt.tera`, with the following content:

```
Hello {{ name }}!
```

If our program has a `name` variable with the value `"Robert"`, we can create a `hello.txt` file with the following content:

```
Hello Robert!
```

As you can see, we can easily create HTML pages with Tera templates. In Tera, there are three delimiters we can use:

- `{{  }}` for **expressions**
- `{%  %}` for **statements**
- `{#  #}` for **comments**

Suppose we have a template named `hello.html.tera` with the following content:

```
<div>
    {# we are setting a variable 'name' with the value
    "Robert" #}
    {% set name = "Robert" %}
    Hello {{ name }}!
</div>
```

We can render that template into a `hello.html` file with the following content:

```
<div>
    Hello Robert!
</div>
```

Tera also has other capabilities such as embedding other templates in a template, basic data operation, control structures, and functions. Basic data operations include basic mathematic operations, basic comparison functions, and string concatenations. Control structures include `if` branching, `for` loops, and other templates. Functions are defined procedures that return some text to be used in the template.

We are going to learn more about some of those capabilities by changing the `OurApplication` responses to use the Tera template engine. Let's set up `OurApplication` to use the Tera template engine:

1. In the `Cargo.toml` file, add the dependencies. We need the `rocket_dyn_templates` crate and the `serde` crate to serialize instances:

```
chrono = {version = "0.4", features = ["serde"]}
rocket_dyn_templates = {path = "../../../rocket/contrib/
dyn_templates/", features = ["tera"]}
serde = "1.0.130"
```

2. Next, add a new configuration in `Rocket.toml` to designate a folder in which to place the template files:

```
[default]
...
template_dir = "src/views"
```

3. In `src/main.rs`, add the following lines to attach the `rocket_dyn_templates::Template` fairing to the application:

```
use rocket_dyn_templates::Template;
...
#[launch]
async fn rocket() -> Rocket<Build> {
...
rocket::build()
...
    .attach(Template::fairing())
```

```
    . . .
    }
```

Adding the `Template` fairing is straightforward. We are going to learn about `Template` in the next section by replacing `RawHtml` with `Template`.

Showcasing users

We are going to modify routes for `/users/<uuid>` and `/users/` by taking the following steps:

1. The first thing we need to do is to create a template. We already configured the folder for templates in `/src/views`, so create a `views` folder in the `src` folder and then, inside the `views` folder, create a template file named `template.html.tera`.

2. We are going to use the file as the base HTML template for all HTML files. Inside `src/views/template.html.tera`, add HTML tags as follows:

```
<!DOCTYPE html>
<html lang="en">
<head>
  <meta charset="utf-8" />
  <title>Our Application User</title>
  <link href="/assets/mini-default.css"
  rel="stylesheet">
  <link rel="icon" type="image/png" href="/assets/
  favicon.png">
  <meta name="viewport" content="width=device-width,
  initial-scale=1">
</head>
<body>
  <div class="container"></div>
</body>
</html>
```

Notice that we included a CSS file in the HTML file. You can download the open source CSS file from `https://minicss.org/` and put it inside the `static` folder. Since we already created a route to serve the static file in `/assets/<filename..>`, we can just use the route directly inside the HTML file.

3. For the next step, we need to include a part where we can put the HTML text we want to render and a part where we can insert `flash` messages. Modify `src/views/template.html.tera` as follows:

```
<div class="container">
  {% if flash %}
    <div class="toast" onclick="this.remove()">
      {{ flash | safe }}
    </div>
  {% endif %}
  {% block body %}{% endblock body %}
</div>
```

By default, all variables are rendered escaped to avoid an XSS (Cross-Site Scripting) attack. We added the condition if there's a `flash` variable, we put the variable inside the `{{ flash }}` expression. To let the HTML tag render as it is and not escaped, we can use the `| safe` filter.

Tera has other built-in filters such as `lower`, `upper`, `capitalize`, and many more. For the content, we are using `block` statements. The `block` statements mean we are going to include another template inside the statement. You can also see `block` ends with an `endblock` statement.

4. Tera can render any type that implements Serde's `Serializable` trait. Let's modify `User` and related types to implement the `Serializable` trait. In `src/models/user.rs`, modify the file as follows:

```
use rocket::serde::Serialize;

...

#[derive(Debug, FromRow, FromForm, Serialize)]
pub struct User {

...
```

5. Modify `src/models/user_status.rs` as follows:

```
use rocket::serde::Serialize;

...

#[derive(sqlx::Type, Debug, FromFormField, Serialize)]
#[repr(i32)]
pub enum UserStatus {

...
```

6. Also, modify src/models/our_date_time.rs as follows:

```
use rocket::serde::Serialize;

#[derive(Debug, sqlx::Type, Clone, Serialize)]
#[sqlx(transparent)]
pub struct OurDateTime(pub DateTime<Utc>);
...
```

7. For Pagination, we can derive Serialize as it is, but using a timestamp in an URL does not look good, for example, /users?pagination.next=2021-12-01T13:09:13.060915Z&pagination.limit=2. We can make the pagination URL look better by converting OurDateTime into i64 and vice versa. In src/models/pagination.rs, append the following lines:

```
use rocket::serde::Serialize;
...
#[derive(Serialize)]
pub struct PaginationContext {
    pub next: i64,
    pub limit: usize,
}

impl Pagination {
    pub fn to_context(&self) -> PaginationContext {
        PaginationContext {
            next: self.next.0.timestamp_nanos(),
            limit: self.limit,
        }
    }
}
```

8. Since we are not going to use RawHtml anymore, remove the use rocket::response::content::RawHtml; directive from the src/routes/mod.rs file and modify the file as follows:

```
use rocket_dyn_templates::Template;
...
type HtmlResponse = Result<Template, Status>;
```

9. In `src/routes/user.rs`, remove the `use`
    `rocket::response::content::RawHtml` directive. We are going to add the
    `Template` directive to make the response return `Template`, but we also need help
    from Serde's `Serialize` and `context!` macros to help convert objects into a
    Tera variable. Append the following lines:

    ```
    use rocket::serde::Serialize;

    ...

    use rocket_dyn_templates::{context, Template};
    ```

10. Then, we can modify the `get_user` function. Inside the `get_user` function in
    `src/routes/user.rs`, delete everything related to RawHtml. Delete all the
    lines starting from `let mut html_string = String::from(USER_HTML_`
    `PREFIX);` to `Ok(RawHtml(html_string))`.

11. Then, replace the content of the deleted lines with the following lines:

    ```
    #[derive(Serialize)]
    struct GetUser {
        user: User,
        flash: Option<String>,
    }
    let flash_message = flash.map(|fm| String::from(fm.
    message()));
    let context = GetUser {
        user,
        flash: flash_message,
    };
    Ok(Template::render("users/show", &context))
    ```

Remember that the Tera template can use any Rust type that implements
the `Serializable` trait. We define the `GetUser` struct that derives the
`Serializable` trait. Since the `User` struct already implements `Serializable`,
we can use it as a field inside the `GetUser` struct. After creating a new instance of
`GetUser`, we then tell the application to render the `"users/show"` template file.

12. Since we have told the application that the template name is `"users/show"`, create a new folder named `users` inside `src/views`. Inside the `src/views/users` folder, create a new file, `src/views/users/show.html.tera`. After that, add these lines inside the file:

```
{% extends "template" %}
{% block body %}
  {% include "users/_user" %}
  <a href="/users/edit/{{user.uuid}}" class="
  button">Edit User</a>
  <form accept-charset="UTF-8" action="/users/
  delete/{{user.uuid}}" autocomplete="off"
  method="POST" id="deleteUser"
      class="hidden"></form>
  <button type="submit" value="Submit"
  form="deleteUser">Delete</button>
  <a href="/users" class="button">User List</a>
{% endblock body %}
```

The first statement, `{% extends "template" %}`, means we are extending `src/views/template.html.tera`, which we created earlier. The parent `src/views/template.html.tera` has a statement, `{% block body %}` `{% endblock body %}`, and we tell the Tera engine to override that block with content from the same block in `src/views/users/show.html.tera`.

13. Inside that code, we also see `{% include "users/_user" %}`, so let's create a `src/views/users/_user.html.tera` file and add the following lines:

```
<div class="row">
  <div class="col-sm-3"><mark>UUID:</mark></div>
  <div class="col-sm-9"> {{ user.uuid }}</div>
</div>
<div class="row">
  <div class="col-sm-3"><mark>Username:</mark></div>
  <div class="col-sm-9"> {{ user.username }}</div>
</div>
<div class="row">
  <div class="col-sm-3"><mark>Email:</mark></div>
  <div class="col-sm-9"> {{ user.email }}</div>
```

```
    </div>
    <div class="row">
      <div class="col-sm-3"><mark>
      Description:</mark></div>
      <div class="col-sm-9"> {{ user.description }}</div>
    </div>
    <div class="row">
      <div class="col-sm-3"><mark>Status:</mark></div>
      <div class="col-sm-9"> {{ user.status }}</div>
    </div>
    <div class="row">
      <div class="col-sm-3"><mark>Created At:</mark></div>
      <div class="col-sm-9"> {{ user.created_at }}</div>
    </div>
    <div class="row">
      <div class="col-sm-3"><mark>Updated At:</mark></div>
      <div class="col-sm-9"> {{ user.updated_at }}</div>
    </div>
```

Inside both files, you will see there are many expressions, such as `{{ user.username }}`. These expressions are using the variable that we defined before: `let context = GetUser { user, flash: flash_message, };`. Then, we tell the application to render the template: `Ok(Template::render("users/show", &context))`.

You might be wondering why we split `show.html.tera` and `_user.html.tera`. One benefit of using a template system is that we can reuse a template. We want to reuse the same user HTML in the `get_users` function.

14. Let's modify the `get_users` function inside the `src/routes/user.rs` files. Delete the lines from `let mut html_string = String::from(USER_HTML_PREFIX);` to `Ok(RawHtml(html_string))`. Replace those lines with the following lines:

    ```
    let context = context! {users: users, pagination: new_
    pagination.map(|pg|pg.to_context())};
    Ok(Template::render("users/index", context))
    ```

15. Instead of defining a new struct, such as `GetUser`, we are using the `context!` macro. By using the `context!` macro, we do not need to create a new type to be passed to the template. Now, create a new file named `src/views/users/index.html.tera`, and add the following lines to the file:

```
{% extends "template" %}
{% block body %}
  {% for user in users %}
    <div class="container">
      <div><mark class="tag">{{loop.
      index}}</mark></div>
      {% include "users/_user" %}
      <a href="/users/{{ user.uuid }}" class="
      button">See User</a>
      <a href="/users/edit/{{ user.uuid }}"
      class="button">Edit User</a>
    </div>
  {% endfor %}
  {% if pagination %}
    <a href="/users?pagination.next={{
    pagination.next}}&pagination.limit={{
    pagination.limit}}" class="button">
      Next
    </a>
  {% endif %}
  <a href="/users/new" class="button">New user</a>
{% endblock %}
```

We see two new things here: a `{% for user in users %}...{% endfor %}` statement, which can be used to iterate arrays, and `{{loop.index}}`, to get the current iteration inside the `for` loop.

16. We want to change the `new_user` and `edit_user` functions too, but before that, we want to see `get_user` and `get_users` in action. Since we already changed the `HtmlResponse` alias into `Result<Template, Status>`, we need to convert `Ok(RawHtml(html_string))` in `new_user` and `edit_user` to use a template too. Change `Ok(RawHtml(html_string))` in the `new_user` and `edit_user` functions to `Ok(Template::render("users/tmp", context!()))`, and create an empty `src/views/users/tmp.html.tera` file.

17. Now, we can run the application and check the page that we have improved with CSS:

Figure 8.1 – get_user() rendered

We can see that the template is working along with the correct CSS file that the application served. In the next section, we will also modify the form to use the template.

Working with forms

If we look at the structure of the form for `new_user` and `edit_user`, we can see that both forms are almost the same, with just a few differences. For example, the forms' `action` endpoints are different, as there are two extra fields for `edit_user`: `_METHOD` and `old_password`. To simplify, we can make one template to be used by both functions. Let's look at the steps:

1. Create a template called `src/views/users/form.html.tera`, and insert the following lines:

```
{% extends "template" %}
{% block body %}
  <form accept-charset="UTF-8" action="{{ form_url }}"
  autocomplete="off" method="POST">
```

```
    <fieldset>
    </fieldset>
  </form>
{% endblock %}
```

2. Next, let's add the title to the form by adding a `legend` tag. Put this inside the `fieldset` tag:

```
<legend>{{ legend }}</legend>
```

3. Under the `legend` tag, we can add an extra field if we are editing the user:

```
{% if edit %}
  <input type="hidden" name="_METHOD" value="PUT" />
{% endif %}
```

4. Continuing with the field, add the fields for `username` and `email` as follows:

```
<div class="row">
  <div class="col-sm-12 col-md-3">
    <label for="username">Username:</label>
  </div>
  <div class="col-sm-12 col-md">
    <input name="username" type="text" value="{{
    user.username }}"/>
  </div>
</div>
<div class="row">
  <div class="col-sm-12 col-md-3">
    <label for="email">Email:</label>
  </div>
  <div class="col-sm-12 col-md">
    <input name="email" type="email" value="{{
    user.email }}"/>
  </div>
</div>
```

5. Add a conditional `old_password` field after the `email` field:

```
{% if edit %}
  <div class="row">
    <div class="col-sm-12 col-md-3">
      <label for="old_password">Old password:</label>
    </div>
    <div class="col-sm-12 col-md">
      <input name="old_password" type="password" />
    </div>
  </div>
{% endif %}
```

6. Add the rest of the fields:

```
<div class="row">
  <div class="col-sm-12 col-md-3">
    <label for="password">Password:</label>
  </div>
  <div class="col-sm-12 col-md">
    <input name="password" type="password" />
  </div>
</div>
<div class="row">
  <div class="col-sm-12 col-md-3">
    <label for="password_confirmation">Password
    Confirmation:</label>
  </div>
  <div class="col-sm-12 col-md">
    <input name="password_confirmation" type=
    "password" />
  </div>
</div>
<div class="row">
  <div class="col-sm-12 col-md-3">
    <label for="description">Tell us a little bit more
    about yourself:</label>
  </div>
```

```
<div class="col-sm-12 col-md">
  <textarea name="description">{{ user.description
  }}</textarea>
</div>
</div>
<button type="submit" value="Submit">Submit</button>
```

7. Then, change the labels and fields to show the value (if there is a value):

```
<input name="username" type="text" {% if user %}value="{{
user.username }}"{% endif %} />
...
<input name="email" type="email" {% if user %}value="{{
user.email }}"{% endif %} />
...
<label for="password">{% if edit %}New Password:{% else
%}Password:{% endif %}</label>
...
<textarea name="description">{% if user %}{{ user.
description }}{% endif %}</textarea>
```

8. After we have created the form template, we can modify the new_user and edit_user functions. In new_user, remove the lines from let mut html_string = String::from(USER_HTML_PREFIX); to Ok(Template::render("users/tmp", context!())) to create RawHtml.

    In form.html.tera, we added these variables: form_url, edit, and legend. We also need to convert Option<FlashMessage<'_>> into a String since the default implementation of the Serializable trait by FlashMessage is not human-readable. Add those variables and render the template in the new_user function as follows:

```
let flash_string = flash
    .map(|fl| format!("{}", fl.message()))
    .unwrap_or_else(|| "".to_string());
let context = context! {
    edit: false,
    form_url: "/users",
    legend: "New User",
    flash: flash_string,
```

```
};
Ok(Template::render("users/form", context))
```

9. For edit_user, we can create the same variables, but this time, we know the data for user so we can include user in the context. Delete the lines in the edit_user function in src/routes/user.rs from let mut html_string = String::from(USER_HTML_PREFIX); to Ok(Template::render("users/tmp", context!())).

   Replace those lines with the following code:

```
let flash_string = flash
    .map(|fl| format!("{}", fl.message()))
    .unwrap_or_else(|| "".to_string());
let context = context! {
    form_url: format!("/users/{}",&user.uuid ),
    edit: true,
    legend: "Edit User",
    flash: flash_string,
    user,
};


Ok(Template::render("users/form", context))
```

10. As the final touch, we can remove the USER_HTML_PREFIX and USER_HTML_SUFFIX constants from src/routes/user.rs. We should also remove the src/views/users/tmp.html.tera file since there's no function using that file anymore. And, since we already enclose the flash message inside the <div></div> tag in the template, we can remove the div usage from flash messages. For example, in src/routes/user.rs, we can modify these lines:

```
Ok(Flash::success(
    Redirect::to(format!("/users/{}", user.uuid)),
    "<div>Successfully created user</div>",
))
```

   We can modify them into the following lines:

```
Ok(Flash::success(
    Redirect::to(format!("/users/{}", user.uuid)),
    "Successfully created user",
))
```

One more thing that we can improve for the form is adding a token to secure the application from **cross-site request forgery** (**CSRF**) attacks. We will learn how to secure our form in the next section.

Securing HTML forms from CSRF

One of the most common security attacks is CSRF, where a malicious third party tricks a user into sending a web form with different values than intended. One way to mitigate this attack is by sending a one-time token along with the form content. The web server then checks the token validity to ensure the request comes from the correct web browser.

We can create such a token in a Rocket application by creating a fairing that will generate a token and check the form value sent back. Let's look at the steps:

1. First, we need to add the dependencies for this. We are going to need a `base64` crate to encode and decode binary values into a string. We also need the `secrets` feature from Rocket to store and retrieve private cookies. Private cookies are just like regular cookies, but they are encrypted by the key we configured in the `Rocket.toml` file with `secret_key`.

   For dependencies, we also need to add `time` as a dependency. Add the following lines in `Cargo.toml`:

   ```
   base64 = {version = "0.13.0"}
   . . .
   rocket = {path = "../../../rocket/core/lib/", features =
   ["uuid", "json", "secrets"]}
   . . .
   time = {version = "0.3", features = ["std"]}
   ```

   The steps for preventing CSRF are generating a random byte, storing the random byte in a private cookie, hashing the random byte as a string, and rendering the form template along with the token string. When the user sends the token back, we can retrieve the token from the cookie and compare both.

2. To make a CSRF fairing, add a new module. In `src/fairings/mod.rs`, add the new module:

   ```
   pub mod csrf;
   ```

3. After that, create a file named `src/fairings/csrf.rs` and add the dependencies and constants for the default value for the cookie to store the random bytes:

```
use argon2::{
    password_hash::{
        rand_core::{OsRng, RngCore},
        PasswordHash, PasswordHasher,
        PasswordVerifier, SaltString,
    },
    Argon2,
};
use rocket::fairing::{self, Fairing, Info, Kind};
use rocket::http::{Cookie, Status};
use rocket::request::{FromRequest, Outcome, Request};
use rocket::serde::Serialize;
use rocket::{Build, Data, Rocket};
use time::{Duration, OffsetDateTime};

const CSRF_NAME: &str = "csrf_cookie";
const CSRF_LENGTH: usize = 32;
const CSRF_DURATION: Duration = Duration::hours(1);
```

Then, we can extend Rocket's `Request` with a new method to retrieve the CSRF token. Because `Request` is an external crate, we cannot add another method, but we can overcome this by adding a trait and making the external crate type extend this trait. We cannot extend an external crate with an external trait, but extending an external crate with an internal trait is permissible.

4. We want to create a method to retrieve CSRF tokens from private cookies. Continue with `src/fairings/csrf.rs` by appending the following lines:

```
trait RequestCsrf {
    fn get_csrf_token(&self) -> Option<Vec<u8>>;
}

impl RequestCsrf for Request<'_> {
    fn get_csrf_token(&self) -> Option<Vec<u8>> {
        self.cookies()
```

```
        .get_private(CSRF_NAME)
        .and_then(|cookie| base64::
        decode(cookie.value()).ok())
        .and_then(|raw| {
            if raw.len() >= CSRF_LENGTH {
                Some(raw)
            } else {
                None
            }
        })
    }
}
```

5. After that, we want to add a fairing that retrieves or generates, and stores random bytes if the cookie does not exist. Add a new struct to be managed as a fairing:

```
#[derive(Debug, Clone)]
pub struct Csrf {}

impl Csrf {
    pub fn new() -> Self {
        Self {}
    }
}


#[rocket::async_trait]
impl Fairing for Csrf {
    fn info(&self) -> Info {
        Info {
            name: "CSRF Fairing",
            kind: Kind::Ignite | Kind::Request,
        }
    }

    async fn on_ignite(&self, rocket: Rocket<Build>) -
> fairing::Result {
        Ok(rocket.manage(self.clone()))
```

```
        }
    }
```

6. We want to retrieve the token first, and if the token does not exist, generate random bytes and add the bytes to the private token. Inside the impl Fairing block, add the on_request function:

```
async fn on_request(&self, request: &mut Request<'_>, _:
&mut Data<'_>) {
    if let Some(_) = request.get_csrf_token() {
        return;
    }

    let mut key = vec![0; CSRF_LENGTH];
    OsRng.fill_bytes(&mut key);

    let encoded = base64::encode(&key[..]);
    let expires = OffsetDateTime::now_utc() + CSRF_
    DURATION;
    let mut csrf_cookie = Cookie::new(
    String::from(CSRF_NAME), encoded);
    csrf_cookie.set_expires(expires);
    request.cookies().add_private(csrf_cookie);
}
```

7. We need a request guard to retrieve the token string from the request. Append the following lines:

```
#[derive(Debug, Serialize)]
pub struct Token(String);

#[rocket::async_trait]
impl<'r> FromRequest<'r> for Token {
    type Error = ();

    async fn from_request(request: &'r Request<'_>) ->
    Outcome<Self, Self::Error> {
        match request.get_csrf_token() {
            None => Outcome::Failure((Status::
```

```
                Forbidden, ())),
                Some(token) => Outcome::
                Success(Self(base64::encode(token))),
            }
        }
    }
```

8. We return an HTTP 403 status code if the token is not found. We also need two
    more functions: generating a hash and comparing the token hash with other strings.
    Since we already use argon2 for password hashing, we can reuse the argon2 crate
    for those functions. Append the following lines:

```
impl Token {
    pub fn generate_hash(&self) -> Result<String,
    String> {
        let salt = SaltString::generate(&mut OsRng);
        Argon2::default()
            .hash_password(self.0.as_bytes(), &salt)
            .map(|hp| hp.to_string())
            .map_err(|_| String::from("cannot hash
            authenticity token"))
    }

    pub fn verify(&self, form_authenticity_token:
    &str) -> Result<(), String> {
        let old_password_hash = self.generate_hash()?;
        let parsed_hash = PasswordHash::new(&old_
        password_hash)
            .map_err(|_| String::from("cannot verify
            authenticity token"))?;
        Ok(Argon2::default()
            .verify_password(form_authenticity_
            token.as_bytes(), &parsed_hash)
            .map_err(|_| String::from("cannot verify
            authenticity token"))?)
    }
}
```

9. After we set up the `Csrf` fairing, we can use it in the application. In `src/main.rs`, attach the fairing to the Rocket application:

```
use our_application::fairings::{csrf::Csrf,
db::DBConnection};

...

async fn rocket() -> Rocket<Build> {

...

        .attach(Csrf::new())

...

}
```

10. In `src/models/user.rs`, add a new field to contain the token sent from the form:

```
pub struct NewUser<'r> {

...

    pub authenticity_token: &'r str,

}

...

pub struct EditedUser<'r> {

...

    pub authenticity_token: &'r str,

}
```

11. In `src/views/users/form.html.tera`, add the field to store the token string:

```
<form accept-charset="UTF-8" action="{{ form_url }}"
autocomplete="off" method="POST">
  <input type="hidden" name="authenticity_token"
  value="{{ csrf_token }}"/>

...
```

12. Finally, we can modify `src/routes/user.rs`. Add the `Token` dependency:

```
use crate::fairings::csrf::Token as CsrfToken;
```

13. We can use `CsrfToken` as a request guard, pass the token to the template, and render the template as HTML:

```
pub async fn new_user(flash: Option<FlashMessage<'_>>,
csrf_token: CsrfToken) -> HtmlResponse {
...
    let context = context! {
        ...
        csrf_token: csrf_token,
    };
    ...
}
...
pub async fn edit_user(
    mut db: Connection<DBConnection>, uuid: &str,
    flash: Option<FlashMessage<'_>>, csrf_token:
    CsrfToken) -> HtmlResponse {
...
    let context = context! {
        ...
        csrf_token: csrf_token,
    };
    ...
}
```

14. Modify the `create_user` function to verify the token and return if the hash does not match:

```
pub async fn create_user<'r>(
    ...
    csrf_token: CsrfToken,
) -> Result<Flash<Redirect>, Flash<Redirect>> {
    ...
    let new_user = user_context.value.as_ref().
    unwrap();
    csrf_token
        .verify(&new_user.authenticity_token)
        .map_err(|_| {
```

```
                    Flash::error(
                        Redirect::to("/users/new"),
                        "Something went wrong when creating
                        user",
                    )
            })?;
        ...
    }
```

15. Do the same with the `update_user`, `put_user`, and `patch_user` functions as well:

```
pub async fn update_user<'r>(
    ...
    csrf_token: CsrfToken,
) -> Result<Flash<Redirect>, Flash<Redirect>> {
    ...
    match user_value.method {
        "PUT" => put_user(db, uuid, user_context,
        csrf_token).await,
        "PATCH" => patch_user(db, uuid, user_context,
        csrf_token).await,
        ...
    }
}
...
pub async fn put_user<'r>(
    ...
    csrf_token: CsrfToken,
) -> Result<Flash<Redirect>, Flash<Redirect>> {
    let user_value = user_context.value.as_ref().
    unwrap();
    csrf_token
        .verify(&user_value.authenticity_token)
        .map_err(|_| {
            Flash::error(
                Redirect::to(format!("/users/edit/{}",
```

```
                        uuid)),
                        "Something went wrong when updating
                        user",
                    )
            })?;
        …
    }
    …
    pub async fn patch_user<'r>(
        . . .
        csrf_token: CsrfToken,
    ) -> Result<Flash<Redirect>, Flash<Redirect>> {
        put_user(db, uuid, user_context, csrf_token).await
    }
```

After that, try relaunching the application and sending the form without the token. We should see the application return an HTTP 403 status code. CSRF is one of the most common web attacks, but we have learned how to mitigate the attack by using Rocket features.

Summary

In this chapter, we have learned about three things that are common for a web application. The first one is learning how to make the Rocket application serve static files by using either PathBuf or the FileServer struct.

Another thing we have learned is how to use rocket_dyn_templates to convert a template into a response to the client. We also learned about a template engine, Tera, and the various capabilities of the Tera template engine.

By utilizing static assets and templates, we can easily create modern web applications. In the next chapter, we are going to learn about user posts: text, picture, and video.

9
Displaying Users' Post

In this chapter, we are going to implement displaying user posts. Along with displaying user posts, we are going to learn about **generic data types** and **trait bounds** to group types that behave similarly and so reduce the creation of similar code. We are also going to learn about the most important part of the Rust programming language: the memory model and its terminologies. We are going to learn more about **ownership**, **moving**, **copying**, **cloning**, **borrowing**, and **lifetime**, and how we can implement those in our code.

After completing this chapter, you will understand and implement those concepts in Rust programming. Generic data types and trait bounds are useful to reduce repetitions, while the Rust memory model and concepts are arguably the most distinctive features of the Rust language and make it not only fast but one of the safest programming languages. Those concepts also make people say that Rust has a steep learning curve.

In this chapter, we are going to cover these main topics:

- Displaying posts – text, photo, and video
- Using generic data types and trait bounds
- Learning about ownership and moving
- Borrowing and lifetime

Technical requirements

For this chapter, we have the same technical requirements as the previous chapter. We need a Rust compiler, a text editor, an HTTP client, and a PostgreSQL database server.

You can find the source code for this chapter at `https://github.com/PacktPublishing/Rust-Web-Development-with-Rocket/tree/main/Chapter09`.

Displaying posts – text, photo, and video

In the previous chapters, we implemented user management, including listing, showing, creating, updating, and deleting user entities. Now, we want to do the same with posts. To refresh your memory, we are planning to have `User` posts. The posts can be either text, photos, or videos.

When we implemented the application skeleton, we created a `Post` struct in `src/models/post.rs` with the following content:

```
pub struct Post {
    pub uuid: Uuid,
    pub user_uuid: Uuid,
    pub post_type: PostType,
    pub content: String,
    pub created_at: OurDateTime,
}
```

The plan is to use `post_type` to differentiate a post based on its type and use the `content` field to store the content of the post.

Now that we have rehashed what we wanted to do, let's implement showing the posts:

1. The first thing we want to do is to create a migration file to change the database schema. We want to create a table to store the posts. In the application root folder, run this command:

    ```
    sqlx migrate add create_posts
    ```

2. We should then see a new file in the `migrations` folder named `YYYYMMDDHHMMSS_create_posts.sql` (depending on the current date-time). Edit the file with the following lines:

    ```
    CREATE TABLE IF NOT EXISTS posts
    (
        uuid        UUID PRIMARY KEY,
        user_uuid   UUID NOT NULL,
        post_type   INTEGER NOT NULL DEFAULT 0,
        content     VARCHAR NOT NULL UNIQUE,
        created_at TIMESTAMPTZ NOT NULL DEFAULT CUR-
        RENT_TIMESTAMP,
        FOREIGN KEY (user_uuid) REFERENCES "users" (uuid)
    );
    ```

3. After editing the file, run the migration in the command line to create the database table:

    ```
    sqlx migrate run
    ```

4. We have also created a `DisplayPostContent` trait in `src/traits/mod.rs`, which has the `raw_html()` method. We want to show content in `Post` by converting content to HTML snippets and rendering the snippets in the Tera template. Change the signature of `raw_html()` so we can use `Post` as the source of HTML snippets:

    ```
    fn raw_html(&self) -> String;
    ```

5. Now, we can implement each of the types in `src/models/text_post.rs`, `src/models/photo_post.rs`, and `src/models/video_post.rs`. Start with changing `src/models/text_post.rs`:

```rust
pub struct TextPost(pub Post);

impl DisplayPostContent for TextPost {
    fn raw_html(&self) -> String {
        format!("<p>{}</p>", self.0.content)
    }
}
```

The implementation is very simple, we are just wrapping the `Post` content inside a p HTML tag.

6. Next, modify `src/models/photo_post.rs`:

```rust
pub struct PhotoPost(pub Post);

impl DisplayPostContent for PhotoPost {
    fn raw_html(&self) -> String {
        format!(
            r#"<figure><img src="{}" class="section
            media"/></figure>"#,
            self.0.content
        )
    }
}
```

For `PhotoPost`, we used the `Post` content as the source of the img HTML tag.

7. The last type we modify is `src/models/video_post.rs`:

```rust
pub struct VideoPost(pub Post);

impl DisplayPostContent for VideoPost {
    fn raw_html(&self) -> String {
        format!(
            r#"<video width="320" height="240" con-
            trols>
<source src="{}" type="video/mp4">
```

```
Your browser does not support the video tag.
</video>"#,
            self.0.content
        )
    }
}
```

For `VideoPost`, we are using the `Post` content as the source of the `video` HTML tag.

We need to create templates for the posts. Let's start with a template that will be used in a single post or multiple posts.

8. Create a `posts` folder in the `src/views` folder. Then, create a `_post.html.tera` file inside the `src/views/posts` folder. Add the following lines to the file:

```
<div class="card fluid">
  {{ post.post_html | safe }}
</div>
```

We are wrapping some content inside a `div` and filtering the content as safe HTML.

9. In the `src/views/posts` folder, create a `show.html.tera` file as a template to show a single post. Add the following lines to the file:

```
{% extends "template" %}
{% block body %}
  {% include "posts/_post" %}
  <button type="submit" value="Submit" form="delete-
  Post">Delete</button>
  <a href="/users/{{user.uuid}}/posts" class="but-
  ton">Post List</a>
{% endblock %}
```

10. Create an `index.html.tera` file inside the `src/views/posts` folder to show user posts. Add the following lines:

```
{% extends "template" %}
{% block body %}
  {% for post in posts %}
    <div class="container">
      <div><mark class="tag">{{ loop.index
```

```
    }}</mark></div>
    {% include "posts/_post" %}
    <a href="/users/{{ user.uuid }}/posts/{{
    post.uuid }}" class="button">See Post</a>
  </div>
{% endfor %}
{% if pagination %}
  <a href="/users/{{ user.uuid }}/posts?pagina
  tion.next={{ pagination.next }}&paginat-
  ion.limit={{ pagination.limit }}" class="button">
    Next
  </a>
{% endif %}
<a href="/users/{{ user.uuid }}/posts/new"
class="button">Upload Post</a>
{% endblock %}
```

11. After creating the views, we can implement methods for the `Post` struct to get the data from the database. Modify the `src/models/post.rs` file to include `use` declarations:

```
use super::bool_wrapper::BoolWrapper;
use super::pagination::{Pagination, DEFAULT_LIMIT};
use super::photo_post::PhotoPost;
use super::post_type::PostType;
use super::text_post::TextPost;
use super::video_post::VideoPost;
use crate::errors::our_error::OurError;
use crate::fairings::db::DBConnection;
use crate::traits::DisplayPostContent;
use rocket::form::FromForm;
use rocket_db_pools::sqlx::{FromRow, PgConnection};
use rocket_db_pools::{sqlx::Acquire, Connection};
```

12. We need to derive `FromRow` for the `Post` struct to convert database rows into `Post` instances:

```
#[derive(FromRow, FromForm)]
pub struct Post {
    ...
}
```

13. Create an `impl` block for `Post`:

```
impl Post {}
```

14. Inside the `impl Post` block, we can add functions to query the database and return the `Post` data. As the functions are similar to the `User` functions, you can copy the code for *steps 14* to *17* in the `Chapter09/01DisplayingPost` source code folder. First, we add the `find()` method to get a single post:

```
pub async fn find(connection: &mut PgConnection, uuid:
&str) -> Result<Post, OurError> {
    let parsed_uuid =
    Uuid::parse_str(uuid).map_err(Our
    Error::from_uuid_error)?;
    let query_str = "SELECT * FROM posts WHERE uuid =
    $1";
    Ok(sqlx::query_as::<_, Self>(query_str)
        .bind(parsed_uuid)
        .fetch_one(connection)
        .await
        .map_err(OurError::from_sqlx_error)?)
}
```

15. Add the `find_all()` method:

```
pub async fn find_all(
    db: &mut Connection<DBConnection>,
    user_uuid: &str,
    pagination: Option<Pagination>,
) -> Result<(Vec<Self>, Option<Pagination>), OurError> {
    if pagination.is_some() {
        return Self::find_all_with_pagination(db,
```

```
        user_uuid, &pagination.unwrap()).await;
    } else {
        return Self::find_all_without_pagination(db,
user_uuid).await;
    }
}
```

16. Add the `find_all_without_pagination()` method:

```rust
async fn find_all_without_pagination(
    db: &mut Connection<DBConnection>,
    user_uuid: &str,
) -> Result<(Vec<Self>, Option<Pagination>), OurError> {
    let parsed_uuid =
    Uuid::parse_str(user_uuid).map_err(Our-
    Error::from_uuid_error)?;
    let query_str = r#"SELECT *
FROM posts
WHERE user_uuid = $1
ORDER BY created_at DESC
LIMIT $2"#;
    let connection = db.acquire().await.map_err(Our-
    Error::from_sqlx_error)?;
    let posts = sqlx::query_as::<_, Self>(query_str)
        .bind(parsed_uuid)
        .bind(DEFAULT_LIMIT as i32)
        .fetch_all(connection)
        .await
        .map_err(OurError::from_sqlx_error)?;
    let mut new_pagination: Option<Pagination> = None;
    if posts.len() == DEFAULT_LIMIT {
        let query_str = "SELECT EXISTS(SELECT 1 FROM
        posts WHERE created_at < $1 ORDER BY
        created_at DESC LIMIT 1)";
        let connection = db.acquire().
        await.map_err(OurError::from_sqlx_error)?;
        let exists = sqlx::query_as::<_,
```

```
BoolWrapper>(query_str)
    .bind(&posts.last().unwrap().created_at)
    .fetch_one(connection)
    .await
    .map_err(OurError::from_sqlx_error)?;
if exists.0 {
    new_pagination = Some(Pagination {
        next: posts.last().unwrap()
        .created_at.to_owned(),
        limit: DEFAULT_LIMIT,
    });
}
}
Ok((posts, new_pagination))
}
```

17. Add the `find_all_with_pagination()` method:

```
async fn find_all_with_pagination(
    db: &mut Connection<DBConnection>,
    user_uuid: &str,
    pagination: &Pagination,
) -> Result<(Vec<Self>, Option<Pagination>), OurError> {
    let parsed_uuid =
    Uuid::parse_str(user_uuid).map_err(
    OurError::from_uuid_error)?;
    let query_str = r#"SELECT *
FROM posts
WHERE user_uuid = $1 AND☐created_at < $2
ORDER BY created_at☐DESC
LIMIT $3"#;
    let connection = db.acquire().await.map_err(
    OurError::from_sqlx_error)?;
    let posts = sqlx::query_as::<_, Self>(query_str)
        .bind(&parsed_uuid)
        .bind(&pagination.next)
        .bind(DEFAULT_LIMIT as i32)
```

```
        .fetch_all(connection)
        .await
        .map_err(OurError::from_sqlx_error)?;
    let mut new_pagination: Option<Pagination> = None;
    if posts.len() == DEFAULT_LIMIT {
        let query_str = "SELECT EXISTS(SELECT 1 FROM
        posts WHERE created_at < $1 ORDER BY
        created_at DESC LIMIT 1)";
        let connection = db.
        acquire().await.map_err(
        OurError::from_sqlx_error)?;
        let exists = sqlx::query_as::<_,
        BoolWrapper>(query_str)
            .bind(&posts.last().unwrap().created_at)
            .fetch_one(connection)
            .await
            .map_err(OurError::from_sqlx_error)?;
        if exists.0 {
            new_pagination = Some(Pagination {
                next: posts.last().unwrap().
                created_at.to_owned(),
                limit: DEFAULT_LIMIT,
            });
        }
    }
    Ok((posts, new_pagination))
}
```

18. We need to add methods to convert a `Post` instance into `TextPost`, `PhotoPost`, or `VideoPost`. Add the following lines inside the `impl Post` block:

```
pub fn to_text(self) -> TextPost {
    TextPost(self)
}
pub fn to_photo(self) -> PhotoPost {
    PhotoPost(self)
}
pub fn to_video(self) -> VideoPost {
    VideoPost(self)
}
```

19. When the view and model implementations are ready, we can implement the function for showing user posts. In `src/routes/post.rs`, add the required use declarations:

```
use crate::models::{pagination::Pagination, post::Post,
post_type::PostType, user::User};
use crate::traits::DisplayPostContent;
use rocket::http::Status;
use rocket::serde::Serialize;
use rocket_db_pools::{sqlx::Acquire, Connection};
use rocket_dyn_templates::{context, Template};
```

20. Modify the `get_post()` function inside `src/routes/post.rs`:

```
#[get("/users/<user_uuid>/posts/<uuid>", format = "text/
html")]
pub async fn get_post(
    mut db: Connection<DBConnection>,
    user_uuid: &str,
    uuid: &str,
) -> HtmlResponse {}
```

21. Inside the `get_post()` function, query the `user` information and the `post` information from the database. Write the following lines:

```
let connection = db
    .acquire()
    .await
    .map_err(|_| Status::InternalServerError)?;
let user = User::find(connection, user_uuid)
    .await
    .map_err(|e| e.status)?;
let connection = db
    .acquire()
    .await
    .map_err(|_| Status::InternalServerError)?;
let post = Post::find(connection, uuid).await.map_err(|e|
e.status)?;
if post.user_uuid != user.uuid {
    return Err(Status::InternalServerError);
}
```

22. In `src/views/posts/show.html.tera` and `src/views/posts/_post.html.tera`, we have set two variables: `user` and `post`. We have to add those two variables into the context passed to the template. Append two structs that will be passed to templates:

```
#[derive(Serialize)]
struct ShowPost {
    post_html: String,
}
#[derive(Serialize)]
struct Context {
    user: User,
    post: ShowPost,
}
```

23. And finally, we can pass the `user` and `post` variables into `context`, render the template along with `context`, and return from the function. Append the following lines:

```
let mut post_html = String::new();
    match post.post_type {
        PostType::Text => post_html =
        post.to_text().raw_html(),
        PostType::Photo => post_html =
        post.to_photo().raw_html(),
        PostType::Video => post_html =
        post.to_video().raw_html(),
    }
    let context = Context {
        user,
        post: ShowPost { post_html },
    };
    Ok(Template::render("posts/show", context))
```

24. For the `get_posts()` function in `src/routes/post.rs`, we want to get the `posts` data from the database. Modify the function into the following lines:

```
#[get("/users/<user_uuid>/posts?<pagination>", format =
"text/html")]
pub async fn get_posts(
    mut db: Connection<DBConnection>,
    user_uuid: &str,
    pagination: Option<Pagination>,
) -> HtmlResponse {
    let user = User::find(&mut db,
    user_uuid).await.map_err(|e| e.status)?;
    let (posts, new_pagination) = Post::find_all(&mut
    db, user_uuid, pagination)
        .await
        .map_err(|e| e.status)?;
}
```

25. Now that we have implemented getting the `posts` data, it's time to render those posts as well. Inside the `get_posts()` function, append the following lines:

```
#[derive(Serialize)]
struct ShowPost {
 uuid: String,
 post_html: String,
}
let show_posts: Vec<ShowPost> = posts
    .into_iter()
    .map(|post| {
        let uuid = post.uuid.to_string();
        let mut post_html = String::new();
        match post.post_type {
            PostType::Text => post_html =
            post.to_text().raw_html(),
            PostType::Photo => post_html =
            post.to_photo().raw_html(),
            PostType::Video => post_html =
            post.to_video().raw_html(),
        };
        ShowPost { uuid, post_html }
    })
    .collect();
let context =
    context! {user, posts: show_posts, pagination:
    new_pagination.map(|pg|pg.to_context())};
Ok(Template::render("posts/index", context))
```

Now we have finished the code for `get_post()` and `get_posts()`, it's time to test those two endpoints. Try adding images and videos to a static folder and add an entry in the database. You can find a sample image and video in the static folder in the source code in the GitHub repository for this chapter. Here is an example:

```
post_type|content
---------+---------------------------------------------------------------
        0|Lorem ipsum dolor sit amet, consectetur adipiscing elit.
        1|/assets/443822918_97d2ae0e60.jpg
        2|/assets/clock.mp4
```

Figure 9.1 – Testing the endpoints

When we open a web browser and navigate to the user posts page, we should be able to see something similar to this screenshot:

Figure 9.2 – Example user posts page

We have implemented the functions to show posts, but if we look back at the code, we can see that all three types (Text, Photo, and Video) have the same method because they are all implementing the same interface.

Let's convert those into generic data types and trait bounds in the next section.

Using generic data types and trait bounds

A **generic data type**, **generic type**, or simply, **generic**, is a way for programming languages to be able to apply the same routine to different data types.

For example, we want to create a `multiplication(a, b) -> c {}` function for different data types, `u8` or `f64`. If a language does not have a generic, a programmer might have to implement two different functions, for example, `multiplication_u8(a: u8, b: u8) -> u8` and `multiplication_f64(a: f64, b: f64) -> f64`. Creating two different functions might look simple, but as the application grows in complexity, the branching and figuring out which function to use will be more complex. If a language has a generic, then the problem of multiple functions can be solved by using a single function that can accept `u8` and `f64`.

In the Rust language, we can make a function to use generics by declaring the generics inside angle brackets after the function name as follows:

```
fn multiplication<T>(a: T, b: T) -> T {}
```

We can also use generics in a `struct` or enum definition. Here is an example:

```
struct Something<T>{
    a: T,
    b: T,
}
enum Shapes<T, U> {
    Rectangle(T, U),
    Circle(T),
}
```

We can also use generics inside method definitions. Following `Something<T>`, we can implement the method as follows:

```
impl<T, U> Something<T, U> {
    fn add(&self, T, U) -> T {}
}
```

At compile time, the compiler identifies and changes the generic code into specific code by using the concrete type (`u8` or `f64` in our multiplication example), depending on which type is used. This process is called **monomorphization**. Because of monomorphization, code written using a generic will produce a binary that has the same execution speed as binary generated using specific code.

Now that we have looked at an introduction to generics, let's use generics in our existing application:

1. In the `src/models/post.rs` file, add another method to convert `Post` instances into `media`:

    ```
    pub fn to_media(self) -> Box<dyn DisplayPostContent> {
        match self.post_type {
            PostType::Text => Box::new(self.to_text()),
            PostType::Photo => Box::new(self.to_photo()),
            PostType::Video => Box::new(self.to_video()),
        }
    }
    ```

    We are telling the `to_media()` method to return the type that implemented `DisplayPostContent` and put `TextPost`, `PhotoPost`, or `VideoPost` into the heap.

2. In the `src/routes/post.rs` file, inside the `get_post()` function, and after the `Context` struct declaration, add the following lines:

    ```
    struct Context {
        ...
    }


    fn create_context<T>(user: User, media: T) -> Context {
        Context {
            user,
            post: ShowPost {
                post_html: media.raw_html(),
            },
        }
    }
    ```

    Yes, we can create a function inside another function. The inner function will have local scope and cannot be used outside the `get_post()` function.

3. We need to change the `context` variable from initiating the struct directly, as follows:

    ```
    let context = Context {...};
    ```

We need to change it into using the `create_context()` function:

```
let media = post.to_media();
let context = create_context(user, media);
```

At this point, we can see that `create_context()` can use any type, such as `String` or `u8`, but `String` and `u8` types don't have the `raw_html()` function. The Rust compiler will show an error when compiling the code. Let's fix this problem by using **trait bounds**.

We have defined and implemented traits several times, and we already know that a trait provides consistent behavior for different data types. We defined the `DisplayPostContent` trait in `src/traits/mod.rs`, and every type that implements `DisplayPostContent` has the same method, `raw_html(&self) -> String`.

We can limit the generic type by adding a trait after the generic declaration. Change the `create_context()` function to use trait bounds:

```
fn create_context<T: DisplayPostContent>(user: User, media: T)
-> Context {...}
```

Unfortunately, using `DisplayPostContent` alone is not enough, because the `T` size is not fixed. We can change the function parameters from `media: T` into a `media: &T` reference, as a reference has a fixed size. We also have another problem, as the `DisplayPostContent` size is not known at compile time, so we need to add another bound. Every `T` type is implicitly expected to have a constant size at compile time, implicitly trait bound to `std::marker::Sized`. We can remove the implicit bound by using a special `?Size` syntax.

We can have more than one trait bound and combine them using the + sign. The resulting signature for the `create_context()` function will be as follows:

```
fn create_context<T: DisplayPostContent + ?Sized>(user: User,
media: &T) -> Context {...}
```

Writing multiple trait bounds inside angle brackets (`<>`) can make the function signature hard to read, so there's an alternative syntax for defining trait bounds:

```
fn create_context<T>(user: User, media: &T) -> Context
where T: DisplayPostContent + ?Sized {...}
```

Because we changed the function signature to use a reference, we have to change the function usage as well:

```
let context = create_context(user, &*media);
```

We get `media` object by dereferencing using the `*` sign and referencing `media` again using the `&` sign.

Now, the Rust compiler should be able to compile the code again. We will learn more about reference in the next two sections, but before that, we have to learn about Rust's memory model called ownership and moving.

Learning about ownership and moving

When we instantiate a struct, we create an **instance**. Imagine a struct as being like a template; an instance is created in the memory based on the template and filled with appropriate data.

An instance in Rust has a **scope**; it is created in a function and gets returned. Here is an example:

```
fn something() -> User {
    let user = User::find(...).unwrap();
    user
}
let user = something()
```

If an instance is not returned, then it's removed from memory because it's not used anymore. In this example, the `user` instance will be removed by the end of the function:

```
fn something() {
    let user = User::find(...).unwrap();
    ...
}
```

We can say that an instance has a scope, as mentioned previously. Any resources created inside a scope will be destroyed by the end of the scope in the *reverse order* of their creation.

We can also create a local scope in a routine by using curly brackets, { }. Any instance created inside the scope will be destroyed by the end of the scope. For example, the user scope is within the curly brackets:

```
...
{
    let user = User::find(...).unwrap();
}
...
```

An instance **owns** resources, not only in **stack memory** but also in **heap memory**. When an instance goes out of scope, either because of function exits or curly brackets scope exits, the resource attached to the instance is automatically cleaned *in reverse order of the creation*. This process is called **resource acquisition is initialization (RAII)**.

Imagine that computer memory consists of a stack and a heap:

```
Stack: ☐☐☐☐☐☐☐☐☐☐☐
Heap:  ☐☐☐☐☐☐☐☐☐☐☐
```

An instance owns memory from stack memory:

```
Stack: ☐☒☒☒☐☐☐☐☐☐☐
Heap:  ☐☐☐☐☐☐☐☐☐☐☐
```

Another instance may own memory from the stack and the heap. For example, a string can be a single word or a couple of paragraphs. We cannot say how large a String instance is going to be, so we cannot store all of the information in stack memory; instead, we can store some in stack memory and some in heap memory. This is a simplification of what it looks like:

```
Stack: ☐☒☐☐☐☐☐☐☐☐☐
          ↓
Heap:  ☐☒☒☒☐☐☐☐☐☐
```

In other programming languages, there's a function called a **destructor**, which is a routine executed when an object is removed from the memory. In Rust, there's a similar trait called **Drop**. In order to execute a function when an object destroyed, a type can implement the std::ops::Drop trait. But, most types don't need to implement the Drop trait and are automatically removed from memory when they're out of scope.

In Rust, if we create an instance and set the instance to another instance, it is called **move**. To see why it's called *move*, let's modify our application code. In the `src/routes/post.rs` file, inside the `get_posts()` function, modify it into the following:

```
let show_posts: Vec<ShowPost> = posts
    .into_iter()
    .map(|post| ShowPost {
        post_html: post.to_media().raw_html(),
        uuid: post.uuid.to_string(),
    })
    .collect();
let context = ...
```

If we compile the program, we should see an error similar to the following:

```
error[E0382]: borrow of moved value: `post`
  --> src/routes/post.rs:78:19
   |
76 |             .map(|post| ShowPost {
   |                  ---- move occurs because `post` has type
`models::post::Post`, which does not implement the `Copy` trait
77 |                 post_html: post.to_media().raw_html(),
   |                           ---------- `post` moved due to
this method call
78 |                 uuid: post.uuid.to_string(),
   |                       ^^^^^^^^^^^^^^^^^^^^^ value borrowed
here after move
```

What is moving? Let's go back to the simplification of memory. When an instance is assigned to another instance, some of the second instance is allocated in stack memory:

```
Stack: □▨□□▨□□□□□□
            ↓
Heap:  □▨▨▨□□□□□□□
```

Then, some of the new instance points to old data in the heap:

```
Stack: □▨□□▨□□□□□□
            ↓
Heap:  □▨▨▨□□□□□□□
```

If both instances point to the same heap memory, what happens if the first instance gets dropped? Because of the possibility of invalid data, in Rust, only one instance may have its own resources. The Rust compiler will refuse to compile code that uses an instance that has been moved.

If we look at our code, the to_media() method in Post moved the post instance and put it inside either TextPost, PhotoPost, or VideoPost. As a result, we cannot use the post instance again in post.uuid.to_string() because it has been moved. Right now, we can fix the code by changing the order of the lines:

```
let show_posts: Vec<ShowPost> = posts
    .into_iter()
    .map(|post| ShowPost {
        uuid: post.uuid.to_string(),
        post_html: post.to_media().raw_html(),
    })
    .collect();
```

There's no moving when we use post.uuid.to_string(), so the code should compile.

But, how we can create a **copy** of an instance instead of moving it? If a type implements the std::marker::Copy trait, then when we assign an instance from another instance, it will create a duplicate in the stack. This is the reason why simple types such as u8, which don't require a lot of memory or have a known size, implement the Copy trait. Let's see the illustration of how this code works:

```
let x: u8 = 8;
let y = x;
Stack: □☒□□☒□□□□□□
Heap:  □□□□□□□□□□□
```

A type may automatically derive the Copy trait if all members of that type implement the Copy trait. We also have to derive Clone, because the Copy trait is trait bound by the Clone trait in its definition: pub trait Copy: Clone { }). Here is an example of deriving the Copy trait:

```
#[derive(Copy, Clone)]
struct Circle {
    r: u8,
}
```

However, this example will not work because `String` does not implement `Copy`:

```
#[derive(Copy, Clone)]
pub struct Sheep {
    ...
    pub name: String,
    ...
}
```

This example will work:

```
#[derive(Clone)]
pub struct Sheep {
    ...
    pub name: String,
    ...
}
```

Cloning works by copying the content of the heap memory. For example, let's say we have the preceding code and the following code:

```
let dolly = Sheep::new(...);
```

We can visualize `dolly` as follows:

```
Stack: ☐☒☐☐☐☐☐☐☐☐☐
          ↓
Heap:   ☐☒☒☒☐☐☐☐☐☐☐
```

Let's say we assign another instance from `dolly`, as follows:

```
let debbie = dolly;
```

This is what the memory usage looks like:

```
Stack: ☐☒☐☐☐☐☒☐☐☐☐
          ↓              ↓
Heap:   ☐☒☒☒☐☒☒☒☒☐☐
```

As allocating heap memory is expensive, we can use another way to see the value of an instance: **borrowing**.

Borrowing and lifetime

We have used **references** in our code. A reference is an instance in the stack that points to another instance. Let's recall what an instance memory usage looks like:

```
Stack:  ☐☒☐☐☐☐☐☐☐☐☐☐
            ↓
Heap:   ☐☒☒☒☐☐☐☐☐☐☐
```

A reference is allocated in stack memory, pointing to another instance:

```
Stack:  ☐☒←☒☐☐☐☐☐☐☐☐
            ↓
Heap:   ☐☒☒☒☐☐☐☐☐☐☐
```

Allocating in the stack is cheaper than allocating in the heap. Because of this, using references most of the time is more efficient than cloning. The process of creating a reference is called **borrowing**, as the reference borrows the content of another instance.

Suppose we have an instance named `airwolf`:

```rust
#[derive(Debug)]
struct Helicopter {
    height: u8,
    cargo: Vec<u8>,
}
let mut airwolf = Helicopter {
    height: 0,
    cargo: Vec::new(),
};
airwolf.height = 10;
```

We can create a reference to `airwolf` by using an ampersand (`&`) operator:

```rust
let camera_monitor_a = &airwolf;
```

Borrowing an instance is like a camera monitor; a reference can see the value of the referenced instance, but the reference cannot modify the value. We can have more than one reference, as seen in this example:

```rust
let camera_monitor_a = &airwolf;
let camera_monitor_b = &airwolf;
```

```
...
let camera_monitor_z = &airwolf;
```

What if we want a reference that can modify the value of the instance it referenced? We can create a **mutable reference** by using the &mut operator:

```
let remote_control = &mut airwolf;
remote_control.height = 15;
```

Now, what will happen if we have two remote controls? Well, the helicopter cannot ascend and descend at the same time. In the same way, Rust restricts mutable references and only allows one mutable reference at a time.

Rust also disallows using mutable references along with immutable references because data inconsistency may occur. For example, adding the following lines will not work:

```
let last_load = camera_monitor_a.cargo.last(); // None
remote_control.cargo.push(100);
```

What is the value of last_load? We expected last_load to be None, but the remote control already pushed something to cargo. Because of the data inconsistency problem, the Rust compiler will emit an error if we try to compile the code.

Implementing borrowing and lifetime

Now that we have learned about ownership, moving, and borrowing, let's modify our code to use references.

1. If we look at the current definition for TextPost, PhotoPost, and VideoPost, we can see we are taking ownership of post and moving the post instance into a new instance of TextPost, PhotoPost, or VideoPost. In src/models/ text_post.rs add the following struct:

    ```
    pub struct TextPost(pub Post);
    ```

2. And in src/models/post.rs , add the following function:

    ```
    pub fn to_text(self) -> TextPost { // self is post
    instance
        TextPost(self) // post is moved into TextPost
    instance
    }
    ```

3. We can convert the TextPost field to be a reference to a Post instance. Modify
    src/models/text_post.rs into the following:

    ```
    pub struct TextPost(&Post);
    ```

4. Since we are converting the unnamed field into a private unnamed field, we also
    need an initializer. Append the following lines:

    ```
    impl TextPost {
        pub fn new(post: &Post) -> Self {
            TextPost(post)
        }
    }
    ```

    Since we changed the initialization of TextPost, we also need to change the
    implementation of to_text() and to_media(). In src/models/post.rs,
    change the to_text() method to the following:

    ```
    pub fn to_text(&self) -> TextPost {
        TextPost::new(self)
    }
    ```

    Change the to_media() method to the following:

    ```
    pub fn to_media(self) -> Box<dyn DisplayPostContent> {
        match self.post_type {
            PostType::Text => Box::new((&self).to_text()),
            ...
        }
    }
    ```

5. Let's try compiling the code. We should see an error as follows:

    ```
    error[E0106]: missing lifetime specifier
      --> src/models/text_post.rs:4:21
       |
    4 | pub struct TextPost(&Post);
       |                     ^ expected named lifetime
    parameter
    ```

The reason for this error is that the code needs a **lifetime specifier**. What is a lifetime specifier? Let's see an example of a very simple program:

```
fn main() {
    let x;
    {
        let y = 5;
        x = &y;
    } // y is out of scope
    println!("{}", *x);
}
```

6. Remember, in Rust, any instance is removed automatically after we reach the end of the scope. In the preceding code, y is created inside a scope denoted by curly brackets, {}. When the code reaches the end of the scope, }, the y instance is cleared from the memory. So, what will happen with x? The preceding code will fail to compile because x is not valid anymore. We can fix the code as follows:

```
fn main() {
    let x;
    {
        let y = 5;
        x = &y;
        println!("{}", *x);
    }
}
```

7. Now, let's take a look at our code in `src/models/text_post.rs`:

```
pub struct TextPost(&Post);
```

Because Rust is multithreaded and has a lot of branching, we cannot guarantee that the reference to the Post instance, &Post, can exist for as long as the TextPost instance. What will happen if &Post is already destroyed while the TextPost instance is not destroyed? The solution is that we place a marker called a **lifetime specifier** or **lifetime annotation**. Let's modify the code definition for TextPost as follows:

```
pub struct TextPost<'a>(&'a Post);
```

We are telling the compiler that any instance of Text Post should live as long as the referenced &Post, which indicated by lifetime indicator, 'a. If the compiler finds out that &Post is not living as long as the Text Post instance, it does not compile the program.

The convention for a lifetime specifier is using a small, single letter such as 'a, but there's also a special lifetime specifier, 'static. A 'static lifetime specifier means the data referenced is living as long as the application. For example, we are saying the data referenced by pi will live as long as the application:

```
let pi: &'static f64 = &3.14;
```

8. Let's modify the rest of the application. We have seen how we use a lifetime specifier in the type definition; let's use it in an impl block and method as well. Modify the rest of src/models/text_post.rs into the following:

```
impl<'a> TextPost<'a> {
    pub fn new(post: &'a Post) -> Self {...}
}
impl<'a> DisplayPostContent for TextPost<'a> {...}
```

9. Let's change PhotoPost in src/models/photo_post.rs to use lifetime as well:

```
pub struct PhotoPost<'a>(&'a Post);

impl<'a> PhotoPost<'a> {
    pub fn new(post: &'a Post) -> Self {
        PhotoPost(post)
    }
}


impl<'a> DisplayPostContent for PhotoPost<'a> {...}
```

10. Let's also change VideoPost in src/models/video_post.rs:

```
pub struct VideoPost<'a>(&'a Post);

impl<'a> VideoPost<'a> {
    pub fn new(post: &'a Post) -> Self {
        VideoPost(post)
    }
```

```
        }

        impl<'a> DisplayPostContent for VideoPost<'a> {...}
```

11. And in src/models/post.rs, modify the code as follows:

```
impl Post {
    pub fn to_text(&self) -> TextPost {
        TextPost::new(self)
    }
    pub fn to_photo(&self) -> PhotoPost {
        PhotoPost::new(self)
    }
    pub fn to_video(&self) -> VideoPost {
        VideoPost::new(self)
    }
    pub fn to_media<'a>(&'a self) -> Box<dyn
    DisplayPostContent + 'a> {
        match self.post_type {
            PostType::Photo => Box::new(self.to_photo()),
            PostType::Text => Box::new(self.to_text()),
            PostType::Video => Box::new(self.to_video()),
        }
    }
    ...
}
```

Now, we are using a borrowed Post instance for TextPost, PhotoPost, or VideoPost instances. But, before we end this chapter, let's refactor the code a little bit by following these instructions:

1. We can see the ShowPost struct is duplicated inside get_post() and get_posts(). Add a new struct into src/models/post.rs:

```
use rocket::serde::Serialize;
...
#[derive(Serialize)]
pub struct ShowPost {
    pub uuid: String,
```

```
        pub post_html: String,
    }
```

2. Add a method to convert `Post` into `ShowPost`:

```
impl Post {
    ...
    pub fn to_show_post<'a>(&'a self) -> ShowPost {
        ShowPost {
            uuid: self.uuid.to_string(),
            post_html: self.to_media().raw_html(),
        }
    }
    ...
}
```

3. In `src/routes/post.rs`, add `ShowPost` to a use declaration:

```
use crate::models::{
    pagination::Pagination,
    post::{Post, ShowPost},
    user::User,
};
```

4. Modify the `get_post()` function by removing these lines to remove unnecessary struct declarations and functions:

```
#[derive(Serialize)]
struct ShowPost {
    post_html: String,
}
#[derive(Serialize)]
struct Context {
    user: User,
    post: ShowPost,
}
fn create_context<T: DisplayPostContent + ?Sized>(user:
User, media: &T) -> Context {
    Context {
        user,
```

```
            post: ShowPost {
                post_html: media.raw_html(),
            },
        }
    }
    let media = post.to_media();
    let context = create_context(user, &*media);
```

5. Replace those lines with the `context!` macro:

    ```
    let context = context! { user, post: &(post.to_show_
    post())};
    ```

6. In the `get_posts()` function, remove these lines:

    ```
    #[derive(Serialize)]
    struct ShowPost {
        uuid: String,
        post_html: String,
    }
    let show_posts: Vec<ShowPost> = posts
        .into_iter()
        .map(|post| ShowPost {
            uuid: post.uuid.to_string(),
            post_html: post.to_media().raw_html(),
        })
        .collect();
    ```

    Replace those lines with this line:

    ```
    let show_posts: Vec<ShowPost> = posts.into_iter().
    map(|post| post.to_show_post()).collect();
    ```

7. Also, change the `context` instantiation:

    ```
    let context = context! {user, posts: &show_posts,
    pagination: new_pagination.map(|pg|pg.to_context())};
    ```

8. And finally, remove the unnecessary use declaration. Remove these lines:

    ```
    use crate::traits::DisplayPostContent;
    use rocket::serde::Serialize;
    ```

The implementation of showing posts should be cleaner now we are using the borrowed `Post` instance. There should be no difference in the speed of the application because we are just using the reference of a single instance.

In fact, sometimes it's better to use an owned attribute instead of a reference because there's no significant performance improvement. Using references can be useful in complex applications, high-memory usage applications, or high-performance applications such as gaming or high-speed trading with a lot of data, at the cost of development time.

Summary

In this chapter, we have implemented `get_post()` and `get_posts()` to show `Post` information in a web browser. Along with those implementations, we have learned about reducing code duplication through generics and trait bounds.

We have also learned about the most distinct and important feature of Rust: its memory model. We now know an instance owns a memory block, either in the stack or in both the stack and heap. We have also learned that assigning another instance to an instance means moving ownership unless it's a simple type that implements the `Copy` and/or `Clone` trait. We have also learned about borrowing, the rules of borrowing, and the use of the lifetime specifier to complement moving, copying, and borrowing.

Those rules are some of the most confusing parts of Rust, but those rules are also what make Rust a very safe language while still having the same performance as other system languages such as C or C++. Now that we have implemented showing posts, let's learn how to upload data in the next chapter.

10

Uploading and Processing Posts

In this chapter, we are going to learn how to upload user posts. We will start with the basics of multipart uploads and continue with `TempFile` to store the uploaded files. After uploading the files, we will implement image processing.

The next thing we are going to learn about is improving processing using concurrent programming techniques, asynchronous programming, and multithreading.

In this chapter, we are going to cover these main topics:

- Uploading a text post
- Uploading a photo post
- Processing files asynchronously
- Uploading a video post and process using worker

Technical requirements

For this chapter, we have the usual requirements: a Rust compiler, a text editor, a web browser, and a PostgreSQL database server. Aside from those requirements, we are going to process uploaded video files. Download the **FFmpeg** command line from `https://www.ffmpeg.org/download.html`. FFmpeg is a multimedia framework to process media files. Make sure you can run FFmpeg on the terminal of your operating system.

You can find the source code for this chapter at `https://github.com/PacktPublishing/Rust-Web-Development-with-Rocket/tree/main/Chapter10`.

Uploading a text post

The first thing we want to upload is a text post because it's the simplest type. When we submit the form in HTML, we can specify the `form` tag `enctype` attribute as `text/plain`, `application/x-www-form-urlencoded`, or `multipart/form-data`. We already learned how to process `application/x-www-form-urlencoded` in the Rocket application when we learned how to create a user. We create a struct and derive `FromForm` for that struct. Later, in the route handling function, we set a route attribute, such as `get` or `post`, and assign the struct in the `data` annotation.

The request body for `Content-Type="application/x-www-form-urlencoded"` is simple: the form keys and values are encoded in key-value tuples separated by &, with an equals sign (=) between the key and the value. If the characters sent are not alphanumeric, they're percent-encoded (%). An example of a form request body is shown here:

```
name=John%20Doe&age=18
```

For uploading a file, `Content-Type` is `multipart/form-data`, and the body is different. Suppose we have the following HTTP header:

```
Content-Type: multipart/form-data; boundary=--------------------
--------charactersforboundary123
```

The HTTP body can be as follows:

```
Content-Type: multipart/form-data; boundary=--------------------
--------charactersforboundary123
Content-Disposition: form-data; name="name"

John Doe
```

```
---------------------------charactersforboundary123
Content-Disposition: form-data; name="upload"; filename="file1.
txt"
Content-Type: text/plain

Lorem ipsum dolor sit amet
---------------------------charactersforboundary123
Content-Disposition: form-data; name="other_field"

Other field
```

In Rocket, we can process `multipart/form-data` by using the `multer` crate. Let's try to implement uploading using that crate by following these instructions:

1. Modify our application by adding these crates into the `Cargo.toml` dependencies:

   ```
   multer = "2.0.2"
   tokio-util = "0.6.9"
   ```

2. Add these configurations in `Rocket.toml` to handle the file upload limit and add a temporary directory to store the uploaded files:

   ```
   limits = {"file/avif" = "1Mib", "file/gif" = "1Mib",
   "file/jpg" = "1Mib", "file/jpeg" = "1Mib", "file/png" =
   "1Mib", "file/svg" = "1Mib", "file/webp" = "1Mib", "file/
   webm" = "64Mib", "file/mp4" = "64Mib", "file/mpeg4" =
   "64Mib", "file/mpg" = "64Mib", "file/mpeg" = "64Mib",
   "file/mov" = "64Mib"}
   temp_dir = "/tmp"
   ```

3. Modify `src/views/posts/index.html.tera` to add a form where the user can upload a file. Add the following lines after the pagination block:

   ```
   <form action="/users/{{ user.uuid }}/posts"
   enctype="multipart/form-data" method="POST">
     <fieldset>
       <legend>New Post</legend>
       <div class="row">
         <div class="col-sm-12 col-md-3">
           <label for="upload">Upload file:</label>
         </div>
   ```

```
        <div class="col-sm-12 col-md">
          <input type="file" name="file" accept="
          text/plain">
        </div>
      </div>
      <button type="submit" value="Submit">Submit</
      button>
    </fieldset>
  </form>
```

4. Add the `create()` method for `Post` in the `src/models/post.rs` file. We want a method to save the `Post` data into the database. Add the following lines inside the `impl Post {}` block:

```
pub async fn create(
    connection: &mut PgConnection,
    user_uuid: &str,
    post_type: PostType,
    content: &str,
) -> Result<Self, OurError> {
    let parsed_uuid = Uuid::parse_str(
    user_uuid).map_err(OurError::from_uuid_error)?;
    let uuid = Uuid::new_v4();
    let query_str = r#"INSERT INTO posts
(uuid, user_uuid, post_type, content)
VALUES
($1, $2, $3, $4)
RETURNING *"#;
    Ok(sqlx::query_as::<_, Self>(query_str)
        .bind(uuid)
        .bind(parsed_uuid)
        .bind(post_type)
        .bind(content)
        .fetch_one(connection)
        .await
        .map_err(OurError::from_sqlx_error)?)
}
```

5. We can remove `FromForm` as we will not use the placeholder anymore. Remove these lines from `src/models/post.rs`:

    ```
    use rocket::form::FromForm;

    ...

    #[derive(FromRow, FromForm)]
    pub struct Post {...}
    ```

6. We need the value from the request's `Content-Type` to get a multipart boundary, but Rocket doesn't have a request guard that can do that. Let's create a type that can handle a raw HTTP `Content-Type` header. In `src/lib.rs`, add the following line:

    ```
    pub mod guards;
    ```

    In the `src` folder, create another folder, named `guards`, and then create a `src/guards/mod.rs` file. Inside the file, add the struct to handle the raw HTTP request body:

    ```
    use rocket::request::{FromRequest, Outcome};


    pub struct RawContentType<'r>(pub &'r str);
    ```

7. Implement `FromRequest` for `RawContent` to create a request guard:

    ```
    #[rocket::async_trait]
    impl<'r> FromRequest<'r> for RawContentType<'r> {
        type Error = ();
        async fn from_request(req: &'r rocket::
        Request<'_>) -> Outcome<Self, Self::Error> {
            let header = req.headers().get_one("
            Content-Type").or(Some("")).unwrap();
            Outcome::Success(RawContentType(header))
        }
    }
    ```

8. Rocket will consider the `"/users/delete/<uuid>"` route as conflicting with the `"/users/<user_uuid>/posts"` route. To avoid that problem, we can add `rank` to the route macro. In `src/routes/user.rs`, edit the route macro above the `delete_user_entry_point()` function:

```
#[post("/users/delete/<uuid>", format = "application/x-
www-form-urlencoded", rank = 2)]
pub async fn delete_user_entry_point(...) -> ... {...}
```

9. In `src/routes/post.rs`, add the required `use` declaration to implement the handling of the HTTP multipart request:

```
use crate::guards::RawContentType;
use crate::models::post_type::PostType;
use multer::Multipart;
use rocket::request::FlashMessage;
use rocket::response::{Flash, Redirect};
use rocket::data::{ByteUnit, Data};
```

10. Add a constant to limit the size of the uploaded file:

```
const TEXT_LIMIT: ByteUnit = ByteUnit::Kibibyte(64);
```

11. Let's modify the `get_posts()` function as well to add a `flash` message if the upload fails or is successful:

```
pub async fn get_posts(
    ...
    flash: Option<FlashMessage<'_>>,
) -> HtmlResponse {
    let flash_message = flash.map(|fm|
    String::from(fm.message()));
    ...
    let context = context! {flash: flash_message,...};
    Ok(Template::render("posts/index", context))
}
```

12. Now it's time to implement the `create_post()` function. The first thing we need to do is modify the `post` route macro and function signature:

```
#[post("/users/<user_uuid>/posts", format = "multipart/
form-data", data = "<upload>", rank = 1)]
pub async fn create_post(
    mut db: Connection<DBConnection>,
    user_uuid: &str,
    content_type: RawContentType<'_>,
    upload: Data<'_>,
) -> Result<Flash<Redirect>, Flash<Redirect>> {...}
```

13. Inside the `create_post()` function, add a closure that returns an error. We add a closure to avoid repetition. Add the following lines:

```
let create_err = || {
    Flash::error(
        Redirect::to(format!("/users/{}/posts",
        user_uuid)),
        "Something went wrong when uploading file",
    )
};
```

14. Under the `create_err` definition, continue by getting the boundary from the `content_type` request guard:

```
let boundary = multer::parse_boundary(content_type.0).
map_err(|_| create_err())?;
```

15. For `TextPost`, we just store the content of the text file in the post's `content` field. Let's open the request body, process it as a multipart, and define a new variable to store the content of the body. Append the following lines:

```
let upload_stream = upload.open(TEXT_LIMIT);
let mut multipart = Multipart::new(tokio_
util::io::ReaderStream::new(upload_stream), boundary);
let mut text_post = String::new();
```

16. The next thing we need to do is to iterate the multipart fields. We can iterate multipart fields as follows:

```
while let Some(mut field) = multipart.next_field().await.
map_err(|_| create_err())? {
    let field_name = field.name();
    let file_name = field.file_name();
    let content_type = field.content_type();
    println!(
        "Field name: {:?}, File name: {:?},
        Content-Type: {:?}",
        field_name, file_name, content_type
    );
}
```

As we only have one field in the form, we can just get the content of the first field and put the value in the text_post variable. Append the following lines:

```
while let Some(mut field) = multipart.next_field().await.
map_err(|_| create_err())? {
    while let Some(field_chunk) =
    field.chunk().await.map_err(|_| create_err())? {
        text_post.push_str(std::str::from_utf8(
        field_chunk.as_ref()).unwrap());
    }
}
```

17. Finally, after we get the request body content and assign it to text_post, it's time to store it in the database and return to the posts list page. Append the following lines:

```
let connection = db.acquire().await.map_err(|_| create_
err())?;
Post::create(connection, user_uuid, PostType::Text,
&text_post)
    .await
    .map_err(|_| create_err())?;
Ok(Flash::success(
    Redirect::to(format!("/users/{}/posts",
    user_uuid)),
    "Successfully created post",
))
```

Now, try restarting the application and uploading the text file. You should see the content of the text file on the `posts` list page:

Figure 10.1 – Uploaded text posts

Now that we have implemented uploading and processing text files, it is time to move on to uploading and processing photo files.

Uploading a photo post

Before Rocket *0.5*, uploading multipart files had to be implemented manually, as in the previous section. Starting from Rocket *0.5*, there's a `rocket::fs::TempFile` type that can be used directly to handle uploaded files.

To handle processing image files, we can use an `image` crate. The crate can handle opening and saving various image file formats. The `image` crate also provides ways to manipulate the image.

Websites process uploaded media files such as images for various reasons, including reducing disk usage. Some websites reduce the image quality and encode the uploaded images into file format with a default smaller size. In this example, we are going to convert all uploaded images into JPEG files with 75% quality.

Let's implement uploading image files using the `image` crate and the `TempFile` struct by following these steps:

1. Remove `multer` and `tokio-util` from `Cargo.toml`. Then, add the `image` crate to `Cargo.toml`:

    ```
    image = "0.24.0"
    ```

2. Remove `pub mod guards;` from `src/lib.rs` and then remove the `src/guards` folder.

3. Add a struct to handle uploaded files in `src/models/post.rs`:

    ```
    use rocket::fs::TempFile;

    . . .

    #[derive(Debug, FromForm)]
    pub struct NewPost<'r> {
        pub file: TempFile<'r>,
    }
    ```

4. Modify `src/views/posts/index.html.tera` to include images as accepted files:

    ```
    . . .

    <input type="file" name="file" accept="text/
    plain,image/*">

    . . .
    ```

5. Remove unused `use` declarations, `TEXT_LIMIT` constant, and part of the `create_post()` function from the boundary variable declaration to the multipart iteration block:

    ```
    use crate::guards::RawContentType;
    use multer::Multipart;
    use rocket::data::{ByteUnit, Data};

    . . .

    const TEXT_LIMIT: ByteUnit = ByteUnit::Kibibyte(64);

    . . .
    ```

```
let boundary = multer::parse_boundary(content_type.0).
map_err(|_| create_err())?;
...until
while let Some(mut field) = multipart.next_field().await.
map_err(|_| create_err())? {
...
}
```

6. Add the required use declarations:

```
use crate::models::post::{NewPost, Post, ShowPost};
use image::codecs::jpeg::JpegEncoder;
use image::io::Reader as ImageReader;
use image::{DynamicImage, ImageEncoder};
use rocket::form::Form;
use std::fs::File;
use std::io::{BufReader, Read};
use std::ops::Deref;
use std::path::Path;
```

7. We can use the NewPost struct that we created earlier as a regular FromForm deriving struct. Modify the create_post() function signature:

```
pub async fn create_post<'r>(
    mut db: Connection<DBConnection>,
    user_uuid: &str,
    mut upload: Form<NewPost<'r>>,
) -> Result<Flash<Redirect>, Flash<Redirect>> {...}
```

8. Under the create_err closure declaration, generate a random uuid name for the new name of the uploaded file:

```
let file_uuid = uuid::Uuid::new_v4().to_string();
```

9. Check Content-Type of the uploaded file, and if the Temp File cannot determine it, return an error:

```
if upload.file.content_type().is_none() {
    return Err(create_err());
}
```

10. Find the extension of the uploaded file and create a new filename:

```
let ext = upload.file.content_type().unwrap().
extension().unwrap();
let tmp_filename = format!("/tmp/{}.{}", &file_uuid,
&ext);
```

11. Persist the uploaded file in the temporary location:

```
upload
    .file
    .persist_to(tmp_filename)
    .await
    .map_err(|_| create_err())?;
```

12. Define content and post_type to be saved later:

```
let mut content = String::new();
let mut post_type = PostType::Text;
```

13. Check the media type of the file. We can separate media types into bitmaps and svg files. For now, we are going to process text and images only. We will process videos in the next section. Append the following lines:

```
let mt = upload.file.content_type().unwrap().deref();
if mt.is_text() {
} else if mt.is_bmp() || mt.is_jpeg() || mt.is_png() ||
mt.is_gif() {
} else if mt.is_svg() {
} else {
    return Err(create_err());
}
```

14. We want to process the text first. Create a vector of byte (u8), open and read the file into the vector, and push the vector into the content String we defined previously. Add these lines inside the mt.is_text() block:

```
let orig_path = upload.file.path().unwrap().to_string_
lossy().to_string();
let mut text_content = vec![];
let _ = File::open(orig_path)
    .map_err(|_| create_err())?
```

```
.read(&mut text_content)
.map_err(|_| create_err())?;
content.push_str(std::str::from_utf8(&text_content).
unwrap());
```

15. Next, we want to process the svg file. For this one, we cannot convert it into a JPEG file; we just want to copy the file into a static folder and create an image path of /assets/random_uuid_filename.svg. Append the following lines inside the mt.is_svg() block:

```
post_type = PostType::Photo;
let dest_filename = format!("{}.svg", file_uuid);
content.push_str("/assets/");
content.push_str(&dest_filename);
let dest_path =
Path::new(rocket::fs::relative!("static")).join(&dest_
filename);
upload
    .file
    .move_copy_to(&dest_path)
    .await
    .map_err(|_| create_err())?;
```

16. For bitmap files, we want to convert them into JPEG files. First, we want to define the destination filename. Append the following lines inside the mt.is_bmp() || mt.is_jpeg() || mt.is_png() || mt.is_gif() block:

```
post_type = PostType::Photo;
let orig_path = upload.file.path().unwrap().to_string_
lossy().to_string();
let dest_filename = format!("{}.jpg", file_uuid);
content.push_str("/assets/");
content.push_str(&dest_filename);
```

17. Continuing the bitmap processing, open the file into a buffer and decode the buffer into a binary format that the image crate understands:

```
let orig_file = File::open(orig_path).map_err(|_| create_
err())?;
let file_reader = BufReader::new(orig_file);
let image: DynamicImage = ImageReader::new(file_reader)
```

```
.with_guessed_format()
.map_err(|_| create_err())?
.decode()
.map_err(|_| create_err())?;
```

18. Create a path for the destination file where we want the JPEG result to be, and create a file at that path. Append the following lines:

```
let dest_path =
Path::new(rocket::fs::relative!("static")).join(&dest_
filename);
let mut file_writer = File::create(dest_path).map_err(|_|
create_err())?;
```

19. We then create a JPEG decoder, specify the JPEG quality and the image attributes, and write the binary format into the destination file. Append the following lines:

```
let encoder = JpegEncoder::new_with_quality(&mut file_
writer, 75);
encoder
    .write_image(
        image.as_bytes(),
        image.width(),
        image.height(),
        image.color(),
    )
    .map_err(|_| create_err())?;
```

20. Finally, we can save the post as in the previous section. Change the Post::create() method as follows:

```
Post::create(connection, user_uuid, post_type, &content)
...
```

We have now finished creating the routine to upload and process text and image files using TempFile and the image crate. Unfortunately, this process uses a more traditional programming paradigm that can be improved. Let's learn how to process the files asynchronously in the next section.

Processing files asynchronously

At the beginning of computer development, the available resources were usually limited in some way. For example, an older generation CPU can only execute one thing at a time. This makes computing difficult because computer resources must wait for the execution of tasks sequentially. For example, while the CPU is calculating a number, the user cannot input anything using the keyboard.

Then, people invented operating systems with a **scheduler**, which assigns resources to run tasks. The invention of the scheduler led to the idea of a **thread**. A thread, or operating system thread, is the smallest sequence of program instructions that can be executed independently by the scheduler.

Some modern programming languages can generate applications that spawn multiple threads at the same time, and so are called **multithreaded** applications.

Creating multithreaded applications can be a drawback, as creating a thread allocates various resources, such as a memory stack. In certain applications, such as desktop applications, it's suitable to create multiple threads. But, creating multiple threads can be a problem in other applications, such as web applications, where requests and responses come and go quickly.

There are techniques to overcome this problem in multiple ways. Some languages opt to have **green threads**, or **virtual threads**, where the language runtime manages a single operating system thread and makes the program behave as if it's multithreaded. Some other languages, such as Javascript and Rust, opt to have **async/await**, a syntactic feature that allows execution parts to be suspended and resumed.

In the previous section, we used the Rust standard library to open and write files for image processing. The library itself is called blocking because, it waits until all the files have been loaded or written. That is not efficient because I/O operations are slower than CPU operations, and the thread can be used to do other operations. We can improve the program by using asynchronous programming.

In Rust, we can declare an `async` function as follows:

```
async fn async_task1() {...}
async fn async_task2() {...}
```

Any async function returns the std::future::Future trait. By default, running the function does not do anything. We can use async_task1 and an executor, such as the futures crate, to run the async function. The following code will behave like regular programming:

```
use futures::executor::block_on;


async fn async_task1() {...}


fn main() {
    let wait = async_task1();
    block_on(wait); // wait until async_task1 finish
}
```

We can use .await after the function usage to not block the thread, as follows:

```
async fn combine() {
    async_task1().await;
    async_task2().await;
}
fn main() {
    block_on(combine());
}
```

Or, we can wait for both functions to finish, as in the following:

```
async fn combine2() {
  let t1 = async_task1();
  let t2 = async_task2();
  futures::join!(t1, t2);
}


fn main() {
  block_on(combine2());
}
```

The `futures` crate is very basic; we can use other runtimes that provide an executor and a scheduler and many other functionalities. There are a couple of competing runtimes in the Rust ecosystem, such as `tokio`, `smol`, and `async-std`. We can use those different runtimes together but it's not very efficient, so it's advised to stick with a single runtime. Rocket itself uses `tokio` as the runtime for `async/await`.

We have used `async` functions in code before, so let's now use `async` functions in more depth. Let's convert the previous image processing to use the `async` programming technique by following these steps:

1. Add the crate dependency in `Cargo.toml`:

    ```
    tokio = {version = "1.16", features = ["fs", "rt"]}
    ```

2. If we look at the code that handles uploading, we can see that file-related operations use a standard library, so they are blocking. We want to replace those libraries with Tokio-equivalent `async` libraries. Remove the `use` declaration from `src/routes/post.rs`:

    ```
    use std::fs::File;
    use std::io::{BufReader, Read};
    ```

    Then, add these `use` declarations:

    ```
    use image::error::ImageError;
    use std::io::Cursor;
    use tokio::fs::File;
    use tokio::io::AsyncReadExt;
    ```

3. Replace the content of the `mt.is_text()` block from the standard library into a Tokio-equivalent library. Find the following lines:

    ```
    let _ = File::open(orig_path)
        .map_err(|_| create_err())?
        .read(&mut text_content)
        .map_err(|_| create_err())?;
    ```

    Replace those lines with the following:

    ```
    let _ = File::open(orig_path)
        .await
        .map_err(|_| create_err())?
        .read_to_end(&mut text_content)
        .await
        .map_err(|_| create_err())?;
    ```

4. Next, replace reading the file in the `mt.is_bmp()` || `mt.is_jpeg()` || `mt.is_png()` || `mt.is_gif()` block. Replace synchronous reading of the file and use a Tokio-equivalent file reading functionality. We want to wrap the result in `std::io::Cursor` because `ImageReader` methods require the `std::io::Read` + `std::io:Seek` trait, and `Cursor` is a type that implemented those traits.

    Find the following lines:

    ```
    let orig_file = File::open(orig_path).map_err(|_| create_
    err())?;
    let file_reader = BufReader::new(orig_file);
    ```

    Replace those lines with the following:

    ```
    let orig_file = tokio::fs::read(orig_path).await.map_
    err(|_| create_err())?;
    let read_buffer = Cursor::new(orig_file);
    ```

5. Wrap the image decoding code in `tokio::task::spawn_blocking`. This function allows synchronous code to run inside the Tokio executor. Find the following lines:

    ```
    let image: DynamicImage = ImageReader::new(file_reader)
        .with_guessed_format()
        .map_err(|_| create_err())?
        .decode()
        .map_err(|_| create_err())?
    ```

    Replace them with the following lines:

    ```
    let encoded_result: Result<DynamicImage, ()> =
    tokio::task::spawn_blocking(|| {
        Ok(ImageReader::new(read_buffer)
            .with_guessed_format()
            .map_err(|_| ())?
            .decode()
            .map_err(|_| ())?)
    })
    .await
    .map_err(|_| create_err())?;
    let image = encoded_result.map_err(|_| create_err())?;
    ```

6. Next, we want to wrap the JPEG encoding in `spawn_blocking` as well. We also want to change file writing into a Tokio `async` function. Find the following lines:

```
let dest_path =
Path::new(rocket::fs::relative!("static")).join(&dest_
filename);
let mut file_writer = File::create(dest_path).map_err(|_|
create_err())?;
JpegEncoder::new_with_quality(&mut file_writer, 75)
    .write_image(
        image.as_bytes(),
        image.width(),
        image.height(),
        image.color(),
    )
    .map_err(|_| create_err())?;
```

Replace them with the following lines:

```
let write_result: Result<Vec<u8>, ImageError> =
tokio::task::spawn_blocking(move || {
    let mut write_buffer: Vec<u8> = vec![];
    let mut write_cursor = Cursor::new(&mut
    write_buffer);
    let _ = JpegEncoder::new_with_quality(&mut
    write_cursor, 75).write_image(
        image.as_bytes(),
        image.width(),
        image.height(),
        image.color(),
    )?;
    Ok(write_buffer)
})
.await
.map_err(|_| create_err())?;
let write_bytes = write_result.map_err(|_| create_
err())?;
let dest_path =
Path::new(rocket::fs::relative!("static")).join(&dest_
filename);
```

```
tokio::fs::write(dest_path, &write_bytes)
    .await
    .map_err(|_| create_err())?;
```

Now, we can run the application and try the uploading functionality again. There should be no differences, except it now uses the `async` function. If there are a lot of requests, an asynchronous application should fare better because the application can use the thread to do other tasks while the application deals with long I/O, such as reading from and writing to a database, dealing with network connections, and handling files, for example.

There is one more example where the application uses `tokio::sync::channel` to create another asynchronous channel, and `rayon` (a crate for data parallelism). You can find this example in the source code for this chapter in the `Chapter10/04UploadingPhotoRayon` folder.

In the next section, let's create the handle for uploading videos and processing videos using a worker.

Uploading a video post and process using a worker

In this section, we are going to process an uploaded video. Processing an uploaded video is not a trivial task as it can take a lot of time, so even with the `async` programming technique, the generated response will take a lot of time.

Another technique in programming to solve a long processing time is using message passing. We are going to create another thread to process the video. When a user uploads a video, we will do the following:

1. Generate a path to the temporary file.
2. Mark the path as unprocessed.
3. Store the path to the file in the database.
4. Send a message from the main Rocket thread into the thread for processing the video.
5. Return the response for uploading the video.

If the thread to process the video receives a message, it will find the data from the database, process the file, and mark the post as finished.

If the user requests the `posts` list or posts while it's still being processed, the user will see the loading image. If the user requests the `posts` list or posts after the processing is finished, the user will see the correct video in the web browser.

Rust libraries for video processing are not very mature yet. There are a couple of libraries that can be used to wrap the `ffmpeg` library, but using the `ffmpeg` library is complicated, even if it's used in its own language, the C language. One solution is to use the `ffmpeg-cli` crate, a wrapper for the `ffmpeg` binary.

Follow these instructions to process uploaded video files:

1. We want to add the `ffmpeg-cli` crate and the `flume` crate as dependencies. The `flume` crate works by generating a channel, a producer, and a consumer. There are similar libraries, such as `std::sync::mpsc` or `crossbeam-channel`, which can be used with varying performance and quality. Add the dependencies to `Cargo.toml`:

    ```
    flume = "0.10.10"
    ffmpeg-cli = "0.1"
    ```

2. Change the form to allow uploading video files. Edit `src/views/posts/index.html.tera`:

    ```
    <input type="file" name="file" accept="text/
    plain,image/*,video/*">
    ```

3. Find a placeholder image to show the video is still being processed. There's a `loading.gif` example file in the source code for this section in `Chapter10/05ProcessingVideo/static/loading.gif`.

4. Modify the `raw_html()` method for `VideoPost` in `src/models/video_post.rs` to show the `loading.gif` image if the video is still not processed yet:

    ```
    fn raw_html(&self) -> String {
        if self.0.content.starts_with("loading") {
            return String::from(
                "<figure><img src=\"/assets/loading.gif\"
                class=\"section media\"/></figure>",
            );
        }
        . . .
    }
    ```

5. We want a method for `Post` to update the content and mark it as permanent. Inside the `impl Post{}` block in `src/models/post.rs`, add the following method:

```
pub async fn make_permanent(
    connection: &mut PgConnection,
    uuid: &str,
    content: &str,
) -> Result<Post, OurError> {
    let parsed_uuid = Uuid::parse_str(uuid).map_err(
    OurError::from_uuid_error)?;
    let query_str = String::from("UPDATE posts SET
    content = $1 WHERE uuid = $2 RETURNING *");
    Ok(sqlx::query_as::<_, Self>(&query_str)
        .bind(content)
        .bind(&parsed_uuid)
        .fetch_one(connection)
        .await
        .map_err(OurError::from_sqlx_error))?
}
```

6. We want to create a message that we want to send to the channel. In `src/models/mod.rs`, add a new module:

```
pub mod worker;
```

7. Then, create a new file, `src/models/worker.rs`. Create a new `Message` struct in the file as follows:

```
pub struct Message {
    pub uuid: String,
    pub orig_filename: String,
    pub dest_filename: String,
}
impl Message {
    pub fn new() -> Self {
        Message {
            uuid: String::new(),
            orig_filename: String::new(),
            dest_filename: String::new(),
```

```
      }
    }
  }
```

8. Create a worker that will be executed when a channel receives a message. Add a new module in `src/lib.rs` called `worker`:

    ```
    pub mod workers;
    ```

9. Create a folder named `workers`. Then, create a new file, `src/workers/mod.rs`, and add a new video module:

    ```
    pub mod video;
    ```

10. Create a new file, `src/workers/video.rs`, and add the required `use` declarations:

    ```
    use crate::models::post::Post;
    use crate::models::worker::Message;
    use ffmpeg_cli::{FfmpegBuilder, File, Parameter};
    use sqlx::pool::PoolConnection;
    use sqlx::Postgres;
    use std::process::Stdio;
    use tokio::runtime::Handle;
    ```

11. Add the function signature to process the video as follows:

    ```
    pub fn process_video(connection: &mut
    PoolConnection<Postgres>, wm: Message) -> Result<(), ()>
    {...}
    ```

12. Inside the `process_video()` function, append these lines to prepare the destination file:

    ```
    let mut dest = String::from("static/");
    dest.push_str(&wm.dest_filename);
    ```

13. We want to re-encode all the files into MP4 files and use the x265 codec for the video file destination. Append these lines to build the parameters for the ffmpeg binary:

```
let builder = FfmpegBuilder::new()
    .stderr(Stdio::piped())
    .option(Parameter::Single("nostdin"))
    .option(Parameter::Single("y"))
    .input(File::new(&wm.orig_filename))
    .output(
        File::new(&dest)
            .option(Parameter::KeyValue("vcodec",
            "libx265"))
            .option(Parameter::KeyValue("crf", "28")),
    );
```

14. The next and final step for the worker is to execute the builder. We can make it async too. Append the following lines:

```
let make_permanent = async {
    let ffmpeg = builder.run().await.unwrap();
    let _ = ffmpeg.process.wait_with_output().
    unwrap();
    let mut display_path = String::from("/assets/");
    display_path.push_str(&wm.dest_filename);
    Post::make_permanent(connection, &wm.uuid,
    &display_path).await
};

let handle = Handle::current();
Ok(handle
    .block_on(make_permanent)
    .map(|_| ())
    .map_err(|_| ())?)
```

15. The next thing we want to do is to create a thread to receive and process the message. We can add a new thread after we initialize Rocket in `src/main.rs`. We want to do several things:

   - Initialize a `worker` thread.

   - Initialize a producer (message sender) and a consumer (message receiver).

   - Initialize a database pool.

   - In the `worker` thread, the consumer will obtain a connection from the database pool and process the message.

   Let's start by adding `use` declarations in `src/main.rs`:

   ```
   use our_application::models::worker::Message;
   use our_application::workers::video::process_video;
   use rocket::serde::Deserialize;
   use sqlx::postgres::PgPoolOptions;
   use tokio::runtime::Handle;
   ```

16. Add the structs to get the database configuration from `Rocket.toml` in `src/main.rs` after the `use` declaration:

   ```
   #[derive(Deserialize)]
   struct Config {
       databases: Databases,
   }
   #[derive(Deserialize)]
   struct Databases {
       main_connection: MainConnection,
   }
   #[derive(Deserialize)]
   struct MainConnection {
       url: String,
   }
   ```

17. In the `rocket()` function after `setup_logger()`, initialize the `flume` producer and consumer as follows:

   ```
   let (tx, rx) = flume::bounded::<Message>(5);
   ```

18. Let Rocket manage the `tx` variable. We also want to assign the generated Rocket object into a variable, because we want to get the database configuration. Find these lines:

```
rocket::build()
    .attach(DBConnection::init())
    .attach(Template::fairing())
    .attach(Csrf::new())
    .mount(...)
```

Replace them with the following lines:

```
let our_rocket = rocket::build()
    .attach(DBConnection::init())
    .attach(Template::fairing())
    .attach(Csrf::new())
    .manage(tx)
    .mount(...);
```

19. After we get `our_rocket`, we want to get the database configuration and initialize a new database connection pool for the worker. Append the following lines:

```
let config: Config = our_rocket
    .figment()
    .extract()
    .expect("Incorrect Rocket.toml configuration");
let pool = PgPoolOptions::new()
    .max_connections(5)
    .connect(&config.databases.main_connection.url)
    .await
    .expect("Failed to connect to database");
```

20. Make a thread that will receive and process the message. Also, don't forget that to return `our_rocket` as a `rocket()` signature, we require the `Rocket<Build>` return value. Append the following lines:

```
tokio::task::spawn_blocking(move || loop {
    let wm = rx.recv().unwrap();
    let handle = Handle::current();
    let get_connection = async { (&pool).
    acquire().await.unwrap() };
```

```
        let mut connection = handle.block_on(get_
        connection);
        let _ = process_video(&mut connection, wm);
    });
    our_rocket
```

21. Now, it's time to use the managed `tx` variable to send a message in the `create_post()` route handling function after we create the video. In `src/routes/post.rs`, add the required `use` declarations:

```
use crate::errors::our_error::OurError;
use crate::models::worker::Message;
use flume::Sender;
use rocket::State;
```

22. In the `create_post()` function, retrieve the `Sender<Message>` instance managed by Rocket. Add the `Sender<Message>` instance to the function parameters:

```
pub async fn create_post<'r>(
    . . .
    tx: &State<Sender<Message>>,
)
```

23. Before the `if mt.is_text()` block, append the following variables:

```
let mut wm = Message::new();
let mut is_video = false;
```

24. After the `if mt.is_svg() {}` block, add a new block to initialize a temporary video value and assign the value to the wm variable we have initialized:

```
else if mt.is_mp4() || mt.is_mpeg() || mt.is_ogg() ||
mt.is_mov() || mt.is_webm() {
    post_type = PostType::Video;
    let dest_filename = format!("{}.mp4", file_uuid);
    content.push_str("loading/assets/");
    content.push_str(&dest_filename);
    is_video = true;
    wm.orig_filename = upload
        .file
```

```
            .path()
            .unwrap()
            .to_string_lossy()
            .to_string()
            .clone();
        wm.dest_filename = dest_filename.clone();
    }
```

25. Find the post creation and return value in the following lines:

```
Post::create(connection, user_uuid, post_type, &content)
    .await
    .map_err(|_| create_err())?;
Ok(Flash::success(
    Redirect::to(format!("/users/{}/posts",
    user_uuid)),
    "Successfully created post",
))
```

Modify this into the following lines:

```
Ok(Post::create(connection, user_uuid, post_type,
&content)
    .await
    .and_then(move |post| {
        if is_video {
            wm.uuid = post.uuid.to_string();
            let _ = tx.send(wm).map_err(|_| {
                OurError::new_internal_server_error(
                    String::from("Cannot process
                    message"),
                    None,
                )
            })?;
        }
        Ok(Flash::success(
            Redirect::to(format!("/users/{}/posts",
            user_uuid)),
            "Successfully created post",
```

```
        ))
    })
    .map_err(|_| create_err())?)
```

Now try restarting the application and uploading the video file; notice the loading page. If the video has been processed, the video should be displayed:

Figure 10.2 – Uploaded video post

Message passing is a very useful technique to process long-running jobs. Try using this technique if your application requires heavy processing but you need to return responses quickly.

Some applications use a more advanced application called a **message broker**, which can retry sending messages, schedule sending messages, send messages to multiple applications, and much more. Some well-known message broker applications are RabbitMQ, ZeroMQ, and Redis. There are many cloud services providing message broker services as well, such as Google Cloud Pub/Sub.

Before we complete this chapter, there's one more thing we can do: delete the user post. Try writing the delete_post() function. You can find the sample code in the Chapter10/06DeletingPost folder on GitHub.

Summary

In this chapter, we have learned several things.

The first thing we learned is how to process multipart forms in Rocket. After that, we learned how to use `TempFile` to upload files. Along with uploading photos and videos, we learned how to process the image files and video files.

We learned more about concurrent programming with `async/await` and multithreading. We also covered how to create a thread and pass a message to a different thread.

In the next chapter, we will focus on how to do authentication, authorization, and serving the API from the Rocket application.

11
Securing and Adding an API and JSON

Two of the most important aspects of a web application are authentication and authorization. In this chapter, we are going to learn how to implement simple authentication and authorization systems. After we have created these systems, we are going to learn how to create a simple **Application Programming Interface** (**API**) and how to protect the API endpoint using a **JSON Web Token** (**JWT**).

At the end of this chapter, you will be able to create an authentication system, with functionality such as logging in and logging out and setting access rights for logged-in users. You will also be able to create an API server and know how to secure the API endpoints.

In this chapter, we are going to cover these main topics:

- Authenticating users
- Authorizing users
- Handling JSON
- Protecting the API with a JWT

Technical requirements

For this chapter, we have the usual requirements: a Rust compiler, a text editor, a web browser, and a PostgreSQL database server, along with the FFmpeg command line. We are going to learn about JSON and APIs in this chapter. Install cURL or any other HTTP testing client.

You can find the source code for this chapter at `https://github.com/ PacktPublishing/Rust-Web-Development-with-Rocket/tree/main/ Chapter11`.

Authenticating users

One of the most common tasks of a web application is handling registration and logging in. By logging in, users can tell the web server that they really are who they say they are.

We already created a sign-up system when we implemented CRUD for the user model. Now, let's implement a login system using the existing user model.

The idea for login is simple: the user can fill in their username and password. The application then verifies that the username and password are valid. After that, the application can generate a cookie with the user's information and return the cookie to the web browser. Every time there's a request from the browser, the cookie is sent back from the browser to the server, and we validate the content of the cookie.

To make sure we don't have to implement the cookie for every request, we can create a request guard that validates the cookie automatically if we use the request guard in a route handling function.

Let's start implementing a user login system by following these steps:

1. Create a request guard to handle user authentication cookies. We can organize request guards in the same place to make it easier if we want to add new request guards. In `src/lib.rs`, add a new module:

   ```
   pub mod guards;
   ```

2. Then, create a folder called `src/guards`. Inside `src/guards`, add a file called `src/guards/mod.rs`. Add a new module in this new file:

   ```
   pub mod auth;
   ```

3. After that, create a new file called `src/guards/auth.rs`.

4. Create a struct to handle user authentication cookies. Let's name the struct
    `CurrentUser`. In `src/guards/auth.rs`, add a struct to store the `User`
    information:

```
use crate::fairings::db::DBConnection;
use crate::models::user::User;
use rocket::http::Status;
use rocket::request::{FromRequest, Outcome, Request};
use rocket::serde::Serialize;
use rocket_db_pools::{sqlx::Acquire, Connection};

#[derive(Serialize)]
pub struct CurrentUser {
    pub user: User,
}
```

5. Define a constant that will be used as a key for the cookie to store the user's
    universally unique identifier (**UUID**):

```
pub const LOGIN_COOKIE_NAME: &str = "user_uuid";
```

6. Implement the `FromRequest` trait for `CurrentUser` to make the struct a request
    guard. Add the implementation skeleton as follows:

```
#[rocket::async_trait]
impl<'r> FromRequest<'r> for CurrentUser {
    type Error = ();
    async fn from_request(req: &'r Request<'_>) ->
    Outcome<Self, Self::Error> {
    }
}
```

7. Inside the `from_request` function, define an error that will be returned if
    something goes wrong:

```
let error = Outcome::Failure((Status::Unauthorized, ()));
```

8. Get the cookie from the request, and extract the UUID from the cookie as follows:

```
let parsed_cookie = req.cookies().get_private(LOGIN_
COOKIE_NAME);
if parsed_cookie.is_none() {
    return error;
}
let cookie = parsed_cookie.unwrap();
let uuid = cookie.value();
```

9. We want to get a connection to the database to find the user information. We can obtain another request guard (such as `Connection<DBConnection>`) inside a request guard implementation. Add the following lines:

```
let parsed_db = req.guard::<Connection<DBConnection>>().
await;
if !parsed_db.is_success() {
    return error;
}
let mut db = parsed_db.unwrap();
let parsed_connection = db.acquire().await;
if parsed_connection.is_err() {
    return error;
}
let connection = parsed_connection.unwrap();
```

10. Find and return the user. Add the following lines:

```
let found_user = User::find(connection, uuid).await;
if found_user.is_err() {
    return error;
}
let user = found_user.unwrap();
Outcome::Success(CurrentUser { user })
```

11. Next, we want to implement the login itself. We will create a `like sessions/ new` route to get the page for the login, a `sessions/create` route to send the username and password for login, and a `sessions/delete` route for logging out. Before implementing those routes, let's create a template for the login. In `src/ views`, add a new folder called `sessions`. Then, create a file called `src/views/ sessions/new.html.tera`. Append the following lines into the file:

```
{% extends "template" %}
{% block body %}
  <form accept-charset="UTF-8" action="login"
  autocomplete="off" method="POST">
    <input type="hidden" name="authenticity_token"
    value="{{ csrf_token }}"/>
    <fieldset>
      <legend>Login</legend>
      <div class="row">
        <div class="col-sm-12 col-md-3">
          <label for="username">Username:</label>
        </div>
        <div class="col-sm-12 col-md">
          <input name="username" type="text" value=""
          />
        </div>
      </div>
      <div class="row">
        <div class="col-sm-12 col-md-3">
          <label for="password">Password:</label>
        </div>
        <div class="col-sm-12 col-md">
          <input name="password" type="password" />
        </div>
      </div>
      <button type="submit" value="Submit">
      Submit</button>
    </fieldset>
  </form>
{% endblock %}
```

12. In `src/models/user.rs`, add a struct for the login information:

```
#[derive(FromForm)]
pub struct Login<'r> {
    pub username: &'r str,
    pub password: &'r str,
    pub authenticity_token: &'r str,
}
```

13. Staying in the same file, we want to create a method for the `User` struct to be able to find the user from the database based on the login username information, and verify whether the login password is correct or not. After verifying that the password is correct by using the `update` method, it is time to refactor this. Create a new function to verify passwords:

```
fn verify_password(ag: &Argon2, reference: &str,
password: &str) -> Result<(), OurError> {
    let reference_hash = PasswordHash::new(
    reference).map_err(|e| {
        OurError::new_internal_server_error(
        String::from("Input error"), Some(
        Box::new(e)))
    })?;
    Ok(ag
        .verify_password(password.as_bytes(),
        &reference_hash)
        .map_err(|e| {
            OurError::new_internal_server_error(
                String::from("Cannot verify
                password"),
                Some(Box::new(e)),
            )
        })?)
}
```

14. Change the `update` method from these lines:

```
let old_password_hash = PasswordHash::new(&old_user.
password_hash).map_err(|e| {
    OurError::new_internal_server_error(
```

```
        String::from("Input error"), Some(Box::new(e)))
    })?;
    let argon2 = Argon2::default();
    argon2
        .verify_password(user.old_password.as_bytes(),
        &old_password_hash)
        .map_err(|e| {
            OurError::new_internal_server_error(
                String::from("Cannot confirm old
                password"),
                Some(Box::new(e)),
            )
        })?;
```

And, change it to the following lines:

```
let argon2 = Argon2::default();
verify_password(&argon2, &old_user.password_hash, user.
old_password)?;
```

15. Create a method to find a user based on the login username. Inside the `impl User` block, add the following method:

```
pub async fn find_by_login<'r>(
    connection: &mut PgConnection,
    login: &'r Login<'r>,
) -> Result<Self, OurError> {
    let query_str = "SELECT * FROM users WHERE
    username = $1";
    let user = sqlx::query_as::<_, Self>(query_str)
        .bind(&login.username)
        .fetch_one(connection)
        .await
        .map_err(OurError::from_sqlx_error)?;
    let argon2 = Argon2::default();
    verify_password(&argon2, &user.password_hash,
    &login.password)?;
    Ok(user)
}
```

16. Now, implement routes for handling login. Create a new mod in `src/routes/`
    `mod.rs`:

    ```
    pub mod session;
    ```

    Then, create a new file called `src/routes/session.rs`.

17. In `src/routes/session.rs`, create a route handling function called `new`.
    We want the function to serve the rendered template for the login that we created
    earlier. Add the following lines:

    ```
    use super::HtmlResponse;
    use crate::fairings::csrf::Token as CsrfToken;
    use rocket::request::FlashMessage;
    use rocket_dyn_templates::{context, Template};


    #[get("/login", format = "text/html")]
    pub async fn new<'r>(flash: Option<FlashMessage<'_>>,
    csrf_token: CsrfToken) -> HtmlResponse {
        let flash_string = flash
            .map(|fl| format!("{}", fl.message()))
            .unwrap_or_else(|| "".to_string());
        let context = context! {
            flash: flash_string,
            csrf_token: csrf_token,
        };

        Ok(Template::render("sessions/new", context))
    }
    ```

18. Then, create a new function called `create`. In this function, we want to find the
    user and verify the password with the password hash in the database. If everything
    goes well, set the cookie with the user information. Append the following lines:

    ```
    use crate::fairings::db::DBConnection;
    use crate::guards::auth::LOGIN_COOKIE_NAME;
    use crate::models::user::{Login, User};
    use rocket::form::{Contextual, Form};
    use rocket::http::{Cookie, CookieJar};
    use rocket::response::{Flash, Redirect};
    ```

```
use rocket_db_pools::{sqlx::Acquire, Connection};
...
#[post("/login", format = "application/x-www-form-
urlencoded", data = "<login_context>")]
pub async fn create<'r>(
    mut db: Connection<DBConnection>,
    login_context: Form<Contextual<'r, Login<'r>>>,
    csrf_token: CsrfToken,
    cookies: &CookieJar<'_>,
) -> Result<Flash<Redirect>, Flash<Redirect>> {
    let login_error = || Flash::error(
    Redirect::to("/login"), "Cannot login");
    if login_context.value.is_none() {
        return Err(login_error());
    }
    let login = login_context.value.as_ref().unwrap();
    csrf_token
        .verify(&login.authenticity_token)
        .map_err(|_| login_error())?;
    let connection = db.acquire().await.map_err(|_|
    login_error())?;
    let user = User::find_by_login(connection, login)
        .await
        .map_err(|_| login_error())?;
    cookies.add_private(Cookie::new(LOGIN_COOKIE_NAME,
    user.uuid.to_string()));
    Ok(Flash::success(Redirect::to("/users"), "Login
    successfully"))
}
```

19. Finally, create a function called `delete`. We will use this function as a route for logging out. Append the following lines:

```
#[post("/logout", format = "application/x-www-form-
urlencoded")]
pub async fn delete(cookies: &CookieJar<'_>) ->
Flash<Redirect> {
    cookies.remove_private(
```

```
        Cookie::named(LOGIN_COOKIE_NAME));
        Flash::success(Redirect::to("/users"), "Logout
        successfully")
    }
```

20. Add `session::new`, `session::create`, and `session::delete` into `src/main.rs`:

```
use our_application::routes::{self, post, session, user};
...

async fn rocket() -> Rocket<Build> {
    ...
    routes![
        ...
        session::new,
        session::create,
        session::delete,
    ]
    ...
}
```

21. Now, we can use `CurrentUser` to ensure that only logged-in users can have access to some endpoints in our application. In `src/routes/user.rs`, remove the routine to find the user in the `edit` endpoint. Delete the following lines:

```
pub async fn edit_user(
    mut db: Connection<DBConnection>,
    ...
) -> HtmlResponse {
    let connection = db
        .acquire()
        .await
        .map_err(|_| Status::InternalServerError)?;
    let user = User::find(connection,
    uuid).await.map_err(|e| e.status)?;
    ...
}
```

22. Then, add `CurrentUser` to the route that requires a logged-in user as follows:

```
use crate::guards::auth::CurrentUser;
...
pub async fn edit_user(...
    current_user: CurrentUser,
) -> HtmlResponse {
    ...
    let context = context! {
        form_url: format!("/users/{}", uuid),
        ...
        user: &current_user.user,
        current_user: &current_user,
        ...
    };
    ...
}
...
pub async fn update_user<'r>(...
    current_user: CurrentUser,
) -> Result<Flash<Redirect>, Flash<Redirect>> {
    ...
    match user_value.method {
        "PUT" => put_user(db, uuid, user_context,
        csrf_token, current_user).await,
        "PATCH" => patch_user(db, uuid, user_context,
        csrf_token, current_user).await,
        ...
    }
}
...
pub async fn put_user<'r>(...
    _current_user: CurrentUser,
) -> Result<Flash<Redirect>, Flash<Redirect>> {...}
...
pub async fn patch_user<'r>(...
    current_user: CurrentUser,
```

```
) -> Result<Flash<Redirect>, Flash<Redirect>> {
    put_user(db, uuid, user_context, csrf_token,
    current_user).await
}
...
pub async fn delete_user_entry_point(...
    current_user: CurrentUser,
) -> Result<Flash<Redirect>, Flash<Redirect>> {
    delete_user(db, uuid, current_user).await
}
...
pub async fn delete_user(...
    _current_user: CurrentUser,
) -> Result<Flash<Redirect>, Flash<Redirect>> {...}
```

23. Finally, protect the endpoint in `src/routes/post.rs` as well. Only logged-in users can upload and delete the post, so modify the code into the following:

```
crate::guards::auth::CurrentUser;
...
pub async fn create_post<'r>(...
    _current_user: CurrentUser,
) -> Result<Flash<Redirect>, Flash<Redirect>> {...}
...
pub async fn delete_post(...
    _current_user: CurrentUser,
) -> Result<Flash<Redirect>, Flash<Redirect>> {...}
```

Before we implemented authentication, we could edit and delete any user or post. Now try editing or deleting something without logging in. Then, try logging in and deleting and editing.

One problem still exists: after logging in, users can edit and delete other users' information. We will learn how to prevent this problem by implementing authorization in the next section.

Authorizing users

Authentication and authorization are two of the main concepts of information security. If authentication is a way to prove that an entity is who they say they are, then authorization is a way to give rights to the entity. One entity might be able to modify some resources, one entity might be able to modify all resources, one entity might only be able to see limited resources, and so on.

In the previous section, we implemented authentication concepts such as login and CurrentUser; now it's time to implement authorization. The idea is that we make sure logged-in users can only modify their own information and posts.

Please keep in mind that this example is very simple. In more advanced information security, there are more advanced concepts, such as role-based access control. For example, we can create a role called admin, we can set a certain user as admin, and admin can do everything without restrictions.

Let's try implementing simple authorization by following these steps:

1. Add a simple method for CurrentUser to compare its instance with a UUID. Append the following lines in src/guards/auth.rs:

   ```
   impl CurrentUser {
       pub fn is(&self, uuid: &str) -> bool {
           self.user.uuid.to_string() == uuid
       }
       pub fn is_not(&self, uuid: &str) -> bool {
           !self.is(uuid)
       }
   }
   ```

2. Add a new type of error as well. Add a new method in src/errors/our_ error.rs in the impl OurError {} block:

   ```
   pub fn new_unauthorized_error(debug: Option<Box<dyn
   Error>>) -> Self {
       Self::new_error_with_status(Status::Unauthorized,
       String::from("unauthorized"), debug)
   }
   ```

3. We can check the `CurrentUser` instance on the templates to control the flow of
    the application. For example, if there's no `CurrentUser` instance, we show the link
    to sign up and log in. If there is a `CurrentUser` instance, we show the link to log
    out. Let's modify the Tera template. Edit `src/views/template.html.tera`
    and append the following lines:

```
<body>
  <header>
    <a href="/users" class="button">Home</a>
    {% if current_user %}
      <form accept-charset="UTF-8" action="/logout"
      autocomplete="off" method="POST" id="logout"
      class="hidden"></form>
      <button type="submit" value="Submit" form="
      logout">Logout</button>
    {% else %}
      <a href="/login" class="button">Login</a>
      <a href="/users/new" class="button">Signup</a>
    {% endif %}
  </header>
  <div class="container">
```

4. Edit `src/views/users/index.html.tera` and remove the following line:

```
<a href="/users/new" class="button">New user</a>
```

    Find this line:

```
<a href="/users/edit/{{ user.uuid }}" class="button">Edit
User</a>
```

    Modify it into the following lines:

```
{% if current_user and current_user.user.uuid == user.
uuid %}
    <a href="/users/edit/{{user.uuid}}" class="
    button">Edit User</a>
{% endif %}
```

5. Edit `src/views/users/show.html.tera` and find these lines:

```
<a href="/users/edit/{{user.uuid}}" class="button">Edit
User</a>
```

```
<form accept-charset="UTF-8" action="/users/delete/
{{user.uuid}}" autocomplete="off" method="POST"
id="deleteUser" class="hidden"></form>
<button type="submit" value="Submit"
form="deleteUser">Delete</button>
```

And, surround those lines with conditional checking as follows:

```
{% if current_user and current_user.user.uuid == user.
uuid %}
    <a href...
    ...
    </button>
{% endif %}
```

6. Next, we want to allow upload only for logged-in users. Find the form lines in
    `src/views/posts/index.html.tera`:

```
<form action="/users/{{ user.uuid }}/posts"
enctype="multipart/form-data" method="POST">
...
</form>
```

Surround the form lines with the following conditional:

```
{% if current_user %}
    <form action="/users/{{ user.uuid }}/posts"
enctype="multipart/form-data" method="POST">
    ...
    </form>
{% endif %}
```

7. Now for the final modification for the template. We want only the owner of the
    post to be able to delete the post. Find these lines in `src/views/posts/show.
    html.tera`:

```
<form accept-charset="UTF-8" action="/users/{{user.
uuid}}/posts/delete/{{post.uuid}}" autocomplete="off"
method="POST" id="deletePost" class="hidden"></form>
<button type="submit" value="Submit"
form="deletePost">Delete</button>
```

Surround them with the following lines:

```
{% if current_user and current_user.user.uuid == user.
uuid %}
    <form...

    ...

    </button>
{% endif %}
```

8. Modify the route handling functions to get the value of current_user. Remember, we can wrap a request guard in Option, such as Option<CurrentUser>. When a route handling function fails to get a CurrentUser instance (for example, there is no logged-in user), it will generate a None variant of Option. We can then pass the instance to a template.

Let's convert route handling functions, starting from src/routes/post.rs. Modify the get_post() function as follows:

```
pub async fn get_post(...
    current_user: Option<CurrentUser>,
) -> HtmlResponse {

    ...

    let context = context! {user, current_user, post:
    &(post.to_show_post())};
    Ok(Template::render("posts/show", context))

}
```

9. Let's do the same thing with the get_posts() function. Modify the function as follows:

```
pub async fn get_posts(...
    current_user: Option<CurrentUser>,
) -> HtmlResponse {
    let context = context! {

        ...

        current_user,
    };
    Ok(Template::render("posts/index", context))

}
```

10. One thing we can do to secure the `create_post()` function is to check whether the user uploading the file has the same UUID as `user_uuid` on the URL. This check is to prevent logged-in attackers from doctoring the request and sending false requests. Put the check in the `create_post()` function before we do file manipulation, as follows:

```
pub async fn create_post<'r>(...
    current_user: CurrentUser,
) -> Result<Flash<Redirect>, Flash<Redirect>> {
    ...
    if current_user.is_not(user_uuid) {
        return Err(create_err());
    }
    ...
}
```

11. We can do the same check for the `delete_post()` function in `src/routes/post.rs`. We want to prevent unauthorized users from being able to send doctored requests and delete other people's posts. Modify `delete_post()` as follows:

```
pub async fn delete_post(...
    current_user: CurrentUser,
) -> Result<Flash<Redirect>, Flash<Redirect>> {
    ...
    if current_user.is_not(user_uuid) {
        return Err(delete_err());
    }
    ...
}
```

12. Try restarting the application, logging in, and seeing whether you can delete other people's posts. Try also modifying `src/routes/user.rs` by applying the same principle: getting the `CurrentUser` instance and applying the necessary check, or passing the `CurrentUser` instance to the template. You can find the full code, including protecting user-related routes, at `https://github.com/PacktPublishing/Rust-Web-Development-with-Rocket/tree/main/Chapter11/02Authorization`.

13. One of the most common tasks of a web server is providing APIs, and some APIs must be secured from unwanted usage. We will learn how to serve an API and protect the API endpoint in the next sections.

Handling JSON

One of the common tasks of web applications is handling APIs. APIs can return a lot of different formats, but modern APIs have converged into two common formats: JSON and XML.

Building an endpoint that returns JSON is pretty simple in the Rocket web framework. For handling the request body in JSON format, we can use `rocket::serde::json::Json<T>` as a data guard. The generic T type must implement the `serde::Deserialize` trait or else the Rust compiler will refuse to compile.

For responding, we can do the same thing by responding with `rocket::serde::json::Json<T>`. The generic T type must only implement the `serde::Serialize` trait when used as a response.

Let's see an example of how to handle JSON requests and responses. We want to create a single API endpoint, `/api/users`. This endpoint can receive a JSON body similar to the structure of our_application::models::pagination::Pagination, as follows:

```
{"next":"2022-02-22T22:22:22.222222Z","limit":10}
```

Follow these steps to implement the API endpoint:

1. Implement `serde::Serialize` for `OurError`. Append these lines into `src/errors/our_error.rs`:

   ```
   use rocket::serde::{Serialize, Serializer};
   use serde::ser::SerializeStruct;
   ...
   impl Serialize for OurError {
       fn serialize<S>(&self, serializer: S) ->
       Result<S::Ok, S::Error>
       where
           S: Serializer,
       {
           let mut state = serializer.
           serialize_struct("OurError", 2)?;
           state.serialize_field("status", &self
           .status.code)?;
           state.serialize_field("message", &self
           .message)?;
   ```

```
            state.end()
        }
    }
```

2. We want `Pagination` to derive `Deserialize` and to automatically implement
    the `Deserialize` trait, as `Pagination` will be used in the JSON data guard,
    `Json<Pagination>`. Because `Pagination` contains the `OurDateTime`
    member, `OurDateTime` has to implement the `Deserialize` trait as well. Modify
    `src/models/our_date_time.rs` and add the `Deserialize` derive macro:

    ```
    use rocket::serde::{Deserialize, Serialize};

    ...

    #[derive(Debug, sqlx::Type, Clone, Serialize,
    Deserialize)]
    #[sqlx(transparent)]
    pub struct OurDateTime(pub DateTime<Utc>);
    ```

3. Derive `Serialize` and `Deserialize` for `Pagination`. We also want to derive
    `Serialize` because we want to use `Pagination` as part of the response from the
    `/api/users` endpoint. Modify `src/models/pagination.rs` as follows:

    ```
    use rocket::serde::{Deserialize, Serialize};

    ...

    #[derive(FromForm, Serialize, Deserialize)]
    pub struct Pagination {...}
    ```

4. For the `User` struct, it already derives `Serialize` automatically, so we can use
    it in a vector of `User`. One thing to be fixed is we don't want the password to be
    included in the resulting JSON. Serde has many macros to control how to generate
    serialized data from a struct. Append a single macro that will skip the `password_`
    `hash` field. Modify `src/models/user.rs`:

    ```
    pub struct User {

        ...

        #[serde(skip_serializing)]
        pub password_hash: String,

        ...

    }
    ```

5. We want to return the vector of `User` and `Pagination` as the resulting JSON. We can create a new struct to wrap those in a field. Append the following lines in `src/models/user.rs`:

    ```
    #[derive(Serialize)]
    pub struct UsersWrapper {
        pub users: Vec<User>,
        #[serde(skip_serializing_if = "Option::is_none")]
        #[serde(default)]
        pub pagination: Option<Pagination>,
    }
    ```

    Note that we are skipping the `pagination` field if it's None.

6. Add a new module in `src/routes/mod.rs`:

    ```
    pub mod api;
    ```

    Then, create a new file in `src/routes/api.rs`.

7. In `src/routes/api.rs`, add the usual use declarations, models, errors, and database connection:

    ```
    use crate::errors::our_error::OurError;
    use crate::fairings::db::DBConnection;
    use crate::models::{
        pagination::Pagination,
        user::{User, UsersWrapper},
    };
    use rocket_db_pools::Connection;
    ```

8. Add a use declaration for `rocket::serde::json::Json` as well:

    ```
    use rocket::serde::json::Json;
    ```

9. Add a route handling function definition to get users:

    ```
    #[get("/users", format = "json", data = "<pagination>")]
    pub async fn users(
        mut db: Connection<DBConnection>,
        pagination: Option<Json<Pagination>>,
    ) -> Result<Json<UsersWrapper>, Json<OurError>> {}
    ```

10. Implement the function. In the function, we can get the content of the JSON using the `into_inner()` method as follows:

```
let parsed_pagination = pagination.map(|p| p.into_
inner());
```

11. Find the users. Append the following lines:

```
let (users, new_pagination) = User::find_all(&mut db,
parsed_pagination)
    .await
    .map_err(|_| OurError::new_internal_server_
    error(String::from("Internal Error"), None))?;
```

Because we have implemented the `Serialize` trait for `OurError`, we can return the type automatically.

12. Now, it's time to return `UsersWrapper`. Append the following lines:

```
Ok(Json(UsersWrapper {
    users,
    pagination: new_pagination,
}))
```

13. The last thing to do is to add the route to `src/main.rs`:

```
use our_application::routes::{self, api, post, session,
user};

...

.mount("/", ...)
.mount("/assets", FileServer::from(relative!("static")))
.mount("/api", routes![api::users])
```

14. Try running the application and sending a request to `http://127.0.0.1:8000/api/users`. We can use any HTTP client, but if we're using cURL, it will be as follows:

```
curl -X GET -H "Content-Type: application/json" -d
"{\"next\":\"2022-02-22T22:22:22.222222Z\",\"limit\":1}"
http://127.0.0.1:8000/api/users
```

The application should return something similar to the following output:

```
{"users":[{"uuid":"8faa59d6-1079-424a-8eb9-09ceef196
9c8","username":"example","email":"example@example.
com","description":"example","status":"Inactive","created_
at":"2021-11-06T06:09:09.534864Z","updated_at":"2021-11-06T06:0
9:09.534864Z"}],"pagination":{"next":"2021-11-06T06:09:09.53486
4Z","limit":1}}
```

Now that we have finished creating an API endpoint, let's try securing the endpoint in the next section.

Protecting the API with a JWT

One common task we want to do is protect the API endpoints from unauthorized access. There are a lot of reasons why API endpoints have to be protected, such as wanting to protect sensitive data, conducting financial services, or offering subscription services.

In the web browser, we can protect server endpoints by making a session, assigning a cookie to the session, and returning the session to the web browser, but an API client is not always a web browser. API clients can be mobile applications, other web applications, hardware monitors, and many more. This raises the question, *how can we protect the API endpoint?*

There are a lot of ways to protect the API endpoint, but one industry standard is by using a JWT. According to *IETF RFC7519*, a JWT is a compact, URL-safe means of representing claims to be transferred between two parties. The claims in a JWT can be either JSON objects or special plaintext representations of said JSON objects.

One flow to use a JWT is as follows:

1. The client sends an authentication request to the server.
2. The server responds with a JWT.
3. The client stores the JWT.
4. The client uses the stored JWT to send an API request.
5. The server verifies the JWT and responds accordingly.

Let's try implementing API endpoint protection by following these steps:

1. Append the required libraries in the `Cargo.toml` dependencies section:

    ```
    hmac = "0.12.1"
    jwt = "0.16.0"
    sha2 = "0.10.2"
    ```

2. We want to use a secret token to sign the JWT token. Add a new entry in `Rocket.toml` as follows:

    ```
    jwt_secret = "fill with your own secret"
    ```

3. Add a new state to store a secret for the token. We want to retrieve the secret when the application creates or verifies JWT. Add the following lines in `src/states/mod.rs`:

    ```
    pub struct JWToken {
        pub secret: String,
    }
    ```

4. Modify `src/main.rs` to make the application retrieve the secret from the configuration and manage the state:

    ```
    use our_application::states::JWToken;
    ...
    struct Config {...
        jwt_secret: String,
    }
    ...
    async fn rocket() -> Rocket<Build> {
        ...
        let config: Config = our_rocket...
        let jwt_secret = JWToken {
            secret: String::from(config.jwt_
            secret.clone()),
        };
        let final_rocket = our_rocket.manage(jwt_secret);
        ...
        final_rocket
    }
    ```

5. Make one struct to hold JSON data that is sent for authentication, and another struct to hold JSON data containing the token to be returned to the client. In src/models/user.rs, add the following use declaration:

```
use rocket::serde::{Deserialize, Serialize};
```

Add the following structs:

```
#[derive(Deserialize)]
pub struct JWTLogin<'r> {
    pub username: &'r str,
    pub password: &'r str,
}
#[derive(Serialize)]
pub struct Auth {
    pub token: String,
}
```

6. Implement a method to verify the username and password for JWTLogin. Add the impl block and method:

```
impl<'r> JWTLogin<'r> {
    pub async fn authenticate(
        &self,
        connection: &mut PgConnection,
        secret: &'r str,
    ) -> Result<Auth, OurError> {}
}
```

7. Inside the authenticate() method, add the error closure:

```
let auth_error =
    || OurError::new_bad_request_error(
    String::from("Cannot verify password"), None);
```

8. Then, find the user according to the username and verify the password:

```
let user = User::find_by_login(
    connection,
    &Login {
        username: self.username,
        password: self.password,
```

```
        authenticity_token: "",
    },
)
.await
.map_err(|_| auth_error())?;
verify_password(&Argon2::default(), &user.password_hash,
self.password)?;
```

9. Add the following use declaration:

```
use hmac::{Hmac, Mac};
use jwt::{SignWithKey};
use sha2::Sha256;
use std::collections::BTreeMap;
```

Continue the following inside authenticate to generate a token from the user's UUID and return the token:

```
let user_uuid = &user.uuid.to_string();
let key: Hmac<Sha256> =
    Hmac::new_from_slice(secret.as_bytes()
    ).map_err(|_| auth_error())?;
let mut claims = BTreeMap::new();
claims.insert("user_uuid", user_uuid);
let token = claims.sign_with_key(&key).map_err(|_| auth_
error())?;
Ok(Auth {
    token: token.as_str().to_string(),
})
```

10. Create a function to authenticate. Let's call this function login(). In src/routes/api.rs, add the required use declaration:

```
use crate::models::user::{Auth, JWTLogin, User,
UsersWrapper};
use crate::states::JWToken;
use rocket::State;
use rocket_db_pools::{sqlx::Acquire, Connection};
```

11. Then, add the `login()` function as follows:

```
#[post("/login", format = "json", data = "<jwt_login>")]
pub async fn login<'r>(
    mut db: Connection<DBConnection>,
    jwt_login: Option<Json<JWTLogin<'r>>>,
    jwt_secret: &State<JWToken>,
) -> Result<Json<Auth>, Json<OurError>> {
    let connection = db
        .acquire()
        .await
        .map_err(|_| OurError::new_internal_server_
        error(String::from("Cannot login"), None))?;
    let parsed_jwt_login = jwt_login
        .map(|p| p.into_inner())
        .ok_or_else(|| OurError::new_bad_request_
        error(String::from("Cannot login"), None))?;
    Ok(Json(
        parsed_jwt_login
            .authenticate(connection, &jwt_secret
            .secret)
            .await
            .map_err(|_| OurError::new_internal_
            server_error(String::from("Cannot login"),
            None))?,
    ))
}
```

12. Now that we have created login functionality, the next action is to create a request guard that handles the authorization token in the request header. In `src/guards/auth.rs`, add the following `use` declarations:

```
use crate::states::JWToken;
use hmac::{Hmac, Mac};
use jwt::{Header, Token, VerifyWithKey};
use sha2::Sha256;
use std::collections::BTreeMap;
```

13. Add a new struct for a request guard called `APIUser`:

```
pub struct APIUser {
    pub user: User,
}
```

14. Implement `FromRequest` for `APIUser`. Add the following block:

```
#[rocket::async_trait]
impl<'r> FromRequest<'r> for APIUser {
    type Error = ();
    async fn from_request(req: &'r Request<'_>) ->
    Outcome<Self, Self::Error> {}
}
```

15. Inside `from_request()`, add the closure to return an error:

```
let error = || Outcome::Failure ((Status::Unauthorized,
())));
```

16. Get the token from the request header:

```
let parsed_header = req.headers().get_
one("Authorization");
if parsed_header.is_none() {
    return error();
}
let token_str = parsed_header.unwrap();
```

17. Get the secret from state:

```
let parsed_secret = req.rocket().state::<JWToken>();
if parsed_secret.is_none() {
    return error();
}
let secret = &parsed_secret.unwrap().secret;
```

18. Verify the token and get the user's UUID:

```
let parsed_key: Result<Hmac<Sha256>, _> = Hmac::new_from_
slice(secret.as_bytes());
if parsed_key.is_err() {
```

```
        return error();
    }
    let key = parsed_key.unwrap();
    let parsed_token: Result<Token<Header, BTreeMap<String,
    String>, _>, _> = token_str.verify_with_key(&key);
    if parsed_token.is_err() {
        return error();
    }
    let token = parsed_token.unwrap();
    let claims = token.claims();
    let parsed_user_uuid = claims.get("user_uuid");
    if parsed_user_uuid.is_none() {
        return error();
    }
    let user_uuid = parsed_user_uuid.unwrap();
```

19. Find the user and return the user data:

```
    let parsed_db = req.guard::<Connection<DBConnection>>().
    await;
    if !parsed_db.is_success() {
        return error();
    }
    let mut db = parsed_db.unwrap();
    let parsed_connection = db.acquire().await;
    if parsed_connection.is_err() {
        return error();
    }
    let connection = parsed_connection.unwrap();
    let found_user = User::find(connection, &user_uuid).
    await;
    if found_user.is_err() {
        return error();
    }
    let user = found_user.unwrap();
    Outcome::Success(APIUser { user })
```

20. Finally, add a new protected API endpoint in `src/routes/api.rs`:

```
use crate::guards::auth::APIUser;

...

#[get("/protected_users", format = "json", data =
"<pagination>")]
pub async fn authenticated_users(
    db: Connection<DBConnection>,
    pagination: Option<Json<Pagination>>,
    _authorized_user: APIUser,
) -> Result<Json<UsersWrapper>, Json<OurError>> {
    users(db, pagination).await
}
```

21. In `src/main.rs`, add the routes to Rocket:

```
...
.mount("/api", routes![api::users, api::login,
 api::authenticated_users])
...
```

Now, try accessing the new endpoint. Here is an example when using the cURL command line:

```
curl -X GET -H "Content-Type: application/json" \
  http://127.0.0.1:8000/api/protected_users
```

The response will be an error. Now try sending a request to get the access token. Here is an example:

```
curl -X POST -H "Content-Type: application/json" \
  -d "{\"username\":\"example\", \"password\": \"password\"}" \
  http://127.0.0.1:8000/api/login
```

There's a token returned, as shown in this example:

```
{"token":"eyJhbGciOiJIUzI1NiJ9.
eyJ1c2VyX3V1aWQiOiJmMGMyZDM4Yy0zNjQ5LTRkOWQtYWQ4My0wZGE4ZmZlY2
E2MDgifQ.XJIaKlIfrBEUw_Ho2HTxd7hQkowTzHkx2q_xKy8HMKA"}
```

Use the token to send the request, as in this example:

```
curl -X GET -H "Content-Type: application/json" \T -H "Content-
Type: application/json" \
 -H "Authorization: eyJhbGciOiJIUzI1NiJ9.
eyJlc2VyX3VlaWQiOiJmMGMyZDM4Yy0zNjQ5LTRkOWQtYWQ4My0wZGE4ZmZlY2
E2MDgifQ.XJIaKlIfrBEUw_Ho2HTxd7hQkowTzHkx2q_xKy8HMKA" \
 http://127.0.0.1:8000/api/protected_users
```

Then, the correct response will be returned. JWT is a good way to protect API endpoints, so use the technique that we have learned when necessary.

Summary

In this chapter, we learned about authenticating users and then creating a cookie to store logged-in user information. We also introduced `CurrentUser` as a request guard that works as authorization for certain parts of the application.

After creating authentication and authorization systems, we also learned about API endpoints. We parsed the incoming request body as a request guard in an API and then created an API response.

Finally, we learned a little bit about the JWT and how to use it to protect API endpoints.

In the next chapter, we are going to learn how to test the code that we have created.

Part 3:
Finishing the Rust
Web Application
Development

In this part, you will learn which parts are not the main part of the Rocket web application, but it's nice to have for Rust and Rocket-related web development.

This part comprises the following chapters:

12

Testing Your Application

Ensuring that a program runs correctly is an important part of programming. In this chapter, we are going to learn about testing the Rust application. We are going to implement a simple unit test for a function, and a functional test for creating a user in our application.

We are going to learn a simple technique to debug and find where a problem occurs in our code.

After learning the information in this chapter, you will be able to create a unit test and functional test for Rust and Rocket applications to ensure the applications work as expected. You will also learn how to use a debugger such as gdb or lldb to debug Rust programs.

In this chapter, we are going to cover these main topics:

- Testing the Rust program

- Testing the Rocket application

- Debugging the Rust application

Technical requirements

In this chapter, we are going to do a test and debug, so we need a debugger. Please install gdb, the GNU Debugger (https://www.sourceware.org/gdb/download/), for your operating system.

You can find the source code for this chapter at `https://github.com/PacktPublishing/Rust-Web-Development-with-Rocket/tree/main/Chapter12`.

Testing the Rust program

One important part of programming is testing the application. There are many kinds of tests, such as unit tests (to test a single function or method), functional tests (to test the function of an application), and integration testing (to test various units and functions as a single combined entity). Various tests should be conducted in order to make the application as correct as intended.

In the Rust standard library, there are three macros to use in testing: `assert!`, `assert_eq!`, and `assert_ne!`. The `assert!` macro accepts one or more parameters. The first parameter is any statement that evaluates to Boolean, and the rest is for debugging if the result is not what is expected.

The `assert_eq!` macro compares equality between the first parameter and second parameter, and the rest is for debugging if the result is not what is expected. The `assert_ne!` macro is the opposite of `assert_eq!`; this macro tests the inequality between the first and the second parameters.

Let's see those macros in action. We want to test the `raw_html()` method of the `TextPost` model in `src/models/text_post.rs`. We want to ensure that the result of the method is the string we want to have. Follow these steps to test the method:

1. In `src/models/text_post.rs`, add the following use declarations:

    ```
    use crate::models::our_date_time::OurDateTime;
    use crate::models::post_type::PostType;
    use chrono::{offset::Utc, TimeZone};
    use uuid::Uuid;
    ```

2. In the same `src/models/text_post.rs` file, we want to have a function for testing. To make a function a test function, annotate the function with the `#[test]` attribute. Add the function declaration:

    ```
    #[test]
    fn test_raw_html() {

    }
    ```

3. Inside the function, initialize a `TextPost` instance as follows:

```
let created_at = OurDateTime(Utc.timestamp_
nanos(1431648000000000));
let post = Post {
    uuid: Uuid::new_v4(),
    user_uuid: Uuid::new_v4(),
    post_type: PostType::Text,
    content: String::from("hello"),
    created_at: created_at,
};
let text_post = TextPost::new(&post);
```

4. Add the `assert!` macro to ensure the resulting string is what we want:

```
assert!(
    text_post.raw_html() ==
    String::from("<p>hel1lo</p>"),
    "String is not equal, {}, {}",
    text_post.raw_html(),
    String::from("<p>hello</p>")
);
```

5. Save the file and run the test by running `cargo test` on the Terminal. As we made a mistake in our test code, `"<p>hel1lo</p>"`, the test should fail, as in the following example:

```
$ cargo test
    Compiling our_application v0.1.0 (/workspace/
    rocketbook/Chapter12/01RustTesting)
...
running 1 test
test models::text_post::test_raw_html ... FAILED

failures:

---- models::text_post::test_raw_html stdout ----
thread 'models::text_post::test_raw_html' panicked at
'String is not equal, <p>hello</p>, <p>hello</p>', src/
models/text_post.rs:33:5
```

```
note: run with `RUST_BACKTRACE=1` environment variable to
display a backtrace


failures:
    models::text_post::test_raw_html


test result: FAILED. 0 passed; 1 failed; 0 ignored; 0
measured; 0 filtered out; finished in 0.00s


error: test failed, to rerun pass '--lib'
```

6. Fix "`<p>helllo</p>`" by replacing it with "`<p>hello</p>`". Save the file and run the test again. The test should work fine now:

```
$ cargo test
    Compiling our_application v0.1.0 (/workspace/
    rocketbook/Chapter12/01RustTesting)
...
running 1 test
test models::text_post::test_raw_html ... ok


test result: ok. 1 passed; 0 failed; 0 ignored; 0
measured; 0 filtered out; finished in 0.00s


    Running unittests (target/debug/deps/our_
    application-43a2db5b02032f30)


running 0 tests


test result: ok. 0 passed; 0 failed; 0 ignored; 0
measured; 0 filtered out; finished in 0.00s


    Doc-tests our_application


running 0 tests


test result: ok. 0 passed; 0 failed; 0 ignored; 0
measured; 0 filtered out; finished in 0.00s
```

7. We want to use the `assert_eq!` and `assert_ne!` macros. The `assert_eq!`
    macro is used to check that the first parameter is equal to the second parameter. The
    `assert_ne!` macro is used to make sure that the first parameter is not equal to
    the second parameter. Add the macros in the `test_raw_html()` function to see
    them in action:

    ```
    assert_eq!(text_post.raw_html(), String::from("<p>hello</
    p>"));
    assert_ne!(text_post.raw_html(),
    String::from("<img>hello</img>"));
    ```

8. Run the test again; it should pass. But, if we look at the test output, there are
    warnings, as follows:

    ```
    warning: unused import: `crate::models::our_date_
    time::OurDateTime`
      --> src/models/text_post.rs:1:5
       |
    1  | use crate::models::our_date_time::OurDateTime;
       |     ^^^^^^^^^^^^^^^^^^^^^^^^^^^^^^^^^^^^^^^^^^^
       |
       = note: `#[warn(unused_imports)]` on by default
    ```

9. One of the conventions for unit testing is to create a test module and mark the
    module as a test so it will not be compiled. In `src/models/text_post.rs`, add
    a new module:

    ```
    #[cfg(test)]
    mod tests {
    }
    ```

10. In `src/models/text_post.rs`, remove these unused use declarations:

    ```
    use crate::models::our_date_time::OurDateTime;
    use crate::models::post_type::PostType;
    use chrono::{offset::Utc, TimeZone};
    use uuid::Uuid;
    ```

11. Vice versa, add the required `use` declarations in `src/models/text_post.rs` in the `tests` module:

```
use super::TextPost;
use crate::models::our_date_time::OurDateTime;
use crate::models::post::Post;
use crate::models::post_type::PostType;
use crate::traits::DisplayPostContent;
use chrono::{offset::Utc, TimeZone};
use uuid::Uuid;
```

12. Move the `test_raw_html()` function into the `tests` module. Run `cargo test` again in the Terminal. The tests should pass with no warnings, as in the following example:

```
$ cargo test
    Finished test [unoptimized + debuginfo] target(s)
    in 0.34s
    Running unittests (target/debug/deps/our_
    application-40cf18b02419edd7)

running 1 test
test models::text_post::tests::test_raw_html ... ok

test result: ok. 1 passed; 0 failed; 0 ignored; 0
measured; 0 filtered out; finished in 0.00s


    Running unittests (target/debug/deps/our_
    application-77e614e023a036bf)

running 0 tests

test result: ok. 0 passed; 0 failed; 0 ignored; 0
measured; 0 filtered out; finished in 0.00s


    Doc-tests our_application
```

```
running 0 tests


test result: ok. 0 passed; 0 failed; 0 ignored; 0
measured; 0 filtered out; finished in 0.00s
```

Now that we have learned how to perform unit tests in Rust, we can continue by testing the application with functional testing in the next section.

Testing the Rocket application

Besides putting the test in the src directory, we can create a test in Rust files in the tests directory inside the root directory of the application. When we run cargo test, the command line will look into the tests directory and run any test found there. People usually use tests in the src directory for unit testing and write functional tests in the tests directory.

The Rocket framework provides a rocket::local module, which contains modules, structs, and methods to send requests to the local Rocket application. The main purpose of sending a non-networked request to the local Rocket application is to inspect the response and ensure that the response is what we expected, mainly for testing.

Let's try implementing integration testing for our application by following these steps:

1. In the root directory of the application, add a new directory named tests. Inside the tests directory, create a file named functional_tests.rs.

2. Inside tests/functional_tests.rs, add a new test function as follows:

    ```
    #[test]
    fn some_test() {
        assert_eq!(2, 1 + 1);
    }
    ```

3. After that, save and run cargo test from the command line. The tests should pass, and the cargo test output should show that it was running the test inside the tests directory, as follows:

    ```
    . . .
    Running tests/functional_tests.rs
    . . .
    ```

4. Let's continue with testing the Rocket application. Create a test function named
    `test_rocket`, but since the application is `async`, we need a different test
    annotation, as follows:

    ```
    #[rocket::async_test]
    async fn test_rocket() {

    }
    ```

5. We are going to put the Rocket instance in the
    `rocket::local::asynchronous::Client` instance. Later, we can use the
    `Client` instance to send a request and verify the response. But, one problem is
    that the Rocket initialization is in `src/main.rs`, not in the `our_application`
    library. We can work around this problem by moving the Rocket initialization
    from `src/main.rs` to `src/lib.rs`. Move the code from `src/main.rs`
    to `src/lib.rs` under the `pub mod` declaration, then change any use
    `our_application::` to use `crate::`.

    After that, rename the `rocket()` function to `setup_rocket()`. Also, add
    `pub` in front of the function and remove `#[launch]` from the top of the
    `setup_rocket()` function.

    We want a method to get the database URL, so implement the `get_database_url`
    method for `Config` in `src/lib.rs`:

    ```
    impl Config {
        pub fn get_database_url(&self) -> String {
            self.databases.main_connection.url.clone()
        }
    }
    ```

6. In `src/main.rs`, change the application to use `setup_rocket()`, as follows:

    ```
    use our_application::setup_rocket;
    use rocket::{Build, Rocket};


    #[launch]
    async fn rocket() -> Rocket<Build> {
        setup_rocket().await
    }
    ```

7. Going back to `tests/functional_test.rs`, add the `our_application` library:

    ```
    use our_application;
    ```

    Then, initialize a Rocket instance in `test_rocket()` as follows:

    ```
    let rocket = our_application::setup_rocket().await;
    ```

8. We want to get a database connection to truncate the database table to ensure a clean state for the testing. Add the required `use` declarations:

    ```
    use our_application::Config;
    use rocket_db_pools::sqlx::PgConnection;
    use sqlx::postgres::PgPoolOptions;
    ```

    Then, add the following lines inside the `test_rocket()` function:

    ```
    let config_wrapper = rocket.figment().extract();
    assert!(config_wrapper.is_ok());
    let config: Config = config_wrapper.unwrap();
    let db_url = config.get_database_url();
    let db_wrapper = PgPoolOptions::new()
        .max_connections(5)
        .connect(&db_url)
        .await;
    assert!(db_wrapper.is_ok());
    let db = db_wrapper.unwrap();
    ```

9. We want to truncate the content of the `users` table. We want a method for `User` to remove all data, but the method should only be available for tests. Let's add a trait to extend the `User` model. Add the `use` declaration:

    ```
    use our_application::models::user::User;
    ```

    Add the `ModelCleaner` trait and implement `ModelCleaner` for `User` as follows:

    ```
    #[rocket::async_trait]
    trait ModelCleaner {
        async fn clear_all(connection: &mut PgConnection)
        -> Result<(), String>;
    }

    #[rocket::async_trait]
    ```

```rust
impl ModelCleaner for User {
    async fn clear_all(connection: &mut PgConnection)
    -> Result<(), String> {
        let _ = sqlx::query("TRUNCATE users RESTART
        IDENTITY CASCADE")
            .execute(connection)
            .await
            .map_err(|_| String::from("error
            truncating databasse"))?;
        Ok(())
    }
}
```

10. Continuing in the `test_rocket()` function, append the following lines:

```rust
let pg_connection_wrapper = db.acquire().await;
assert!(pg_connection_wrapper.is_ok());
let mut pg_connection = pg_connection_wrapper.unwrap();
let clear_result_wrapper = User::clear_all(&mut pg_
connection).await;
assert!(clear_result_wrapper.is_ok());
```

11. In the `tests/functional_tests.rs` file, add the `use` declaration:

```rust
use rocket::local::asynchronous::Client;
```

Then, create a `Client` instance inside the `test_rocket()` function:

```rust
let client_wrapper = Client::tracked(rocket).await;
assert!(client_wrapper.is_ok());
let client = client_wrapper.unwrap();
```

12. Right now, the number of users in the database is 0. We want to make a test by getting `"/users"` and parsing the HTML. One crate to parse HTML is `scraper`. Because we only want to use the `scraper` crate for testing, add a new part in `Cargo.toml` called `[dev-dependencies]`, as follows:

```toml
[dev-dependencies]
scraper = "0.12.0"
```

13. Going back to `tests/functional_test.rs`, we want to get the `"/users"` response. Add the `use` declaration:

```
use rocket::http::Status;
```

Then, append the following lines inside the `test_rocket()` function:

```
let req = client.get("/users");
let resp = req.dispatch().await;
assert_eq!(resp.status(), Status::Ok);
```

14. We want to verify that the response body does not contain any users. If we look at the `src/views/users/index.html.tera` template, we see there's a `mark` HTML tag for each user. Let's use `scraper` to verify the response by adding the `use` declaration:

```
use scraper::{Html, Selector};
```

Then, append the following lines inside the `test_rocket()` function:

```
let body_wrapper = resp.into_string().await;
assert!(body_wrapper.is_some());
let body = Html::parse_document(&body_wrapper.unwrap());
let selector = Selector::parse(r#"mark.tag"#).unwrap();
let containers = body.select(&selector);
let num_of_elements = containers.count();
assert_eq!(num_of_elements, 0);
```

15. We want to create a `post` request to create a new user, but one problem is that the application will do an token authenticity check, so we need to get the value from the `"/users/new"` page first. Append the following lines to get the token from the response body:

```
let req = client.get("/users/new");
let resp = req.dispatch().await;
assert_eq!(resp.status(), Status::Ok);
let body_wrapper = resp.into_string().await;
assert!(body_wrapper.is_some());
let body = Html::parse_document(&body_wrapper.unwrap());
let authenticity_token_selector =
Selector::parse(r#"input[name="authenticity_token"]"#).
unwrap();
```

```
let element_wrapper = body.select(&authenticity_token_
selector).next();
assert!(element_wrapper.is_some());
let element = element_wrapper.unwrap();
let value_wrapper = element.value().attr("value");
assert!(value_wrapper.is_some());
let authenticity_token = value_wrapper.unwrap();
```

16. Use `authenticity_token` to send the post request. Add the use declaration:

```
use rocket::http::ContentType;
```

17. Then, append the following lines to the `test_rocket()` function:

```
let username = "testing123";
let password = "lkjKLAJ09231478mlasdfkjsdkj";
let req = client.post("/users")
    .header(ContentType::Form)
    .body(
        format!("authenticity_token={
        }&username={}&email={}@{}
        .com&password={}&password_confirmation={}
        &description=",
    authenticity_token, username, username, username,
    password, password,
));
let resp = req.dispatch().await;
assert_eq!(resp.status(), Status::SeeOther);
```

18. Finally check the `"/users"` page again; you should see one user. Append the following lines:

```
let req = client.get("/users");
let resp = req.dispatch().await;
assert_eq!(resp.status(), Status::Ok);
let body_wrapper = resp.into_string().await;
assert!(body_wrapper.is_some());
let body = Html::parse_document(&body_wrapper.unwrap());
let selector = Selector::parse(r#"mark.tag"#).unwrap();
```

```
let containers = body.select(&selector);
let num_of_elements = containers.count();
assert_eq!(num_of_elements, 1);
```

Try running the test again. Sometimes the test works:

```
$ cargo test
...
      Running tests/functional_tests.rs (target/debug/deps/
functional_tests-625b16e4b25b72de)

running 2 tests
test some_test ... ok
...
test test_rocket ... ok

test result: ok. 2 passed; 0 failed; 0 ignored; 0 measured; 0
filtered out; finished in 2.43s
```

But, sometimes, the test doesn't work:

```
$ cargo test
...
      Running tests/functional_tests.rs (target/
      debug/deps/functional_tests-625b16e4b25b72de)

running 2 tests
test some_test ... ok
...
test test_rocket ... FAILED

failures:

---- test_rocket stdout ----
thread 'test_rocket' panicked at 'assertion failed: `(left ==
right)`
  left: `0`,
 right: `1`', tests/functional_tests.rs:115:5
```

```
note: run with `RUST_BACKTRACE=1` environment variable to
display a backtrace
thread 'rocket-worker-test-thread' panicked at 'called
`Result::unwrap()` on an `Err` value: Disconnected',
/workspace/rocketbook/Chapter12/03RocketTesting/src/lib.
rs:137:28


failures:
    test_rocket


test result: FAILED. 1 passed; 1 failed; 0 ignored; 0 measured;
0 filtered out; finished in 2.09s
```

Why did the test fail? We will learn how to debug the Rust program in the next section.

Debugging the Rust application

In the previous section, we learned about writing functional tests, but sometimes, the test fails. We want to know why the test failed. There are two possible places where the error might occur. One is in the user creation process, and the other is in finding users after creating the user.

One way to debug is by logging where the error might occur. If we log all the possible errors in the user creation process (for example, in `src/routes/user.rs` in the `create_user()` function), we will find out that the authenticity token verification sometimes produces an error. An example of logging the error is as follows:

```
csrf_token
    .verify(&new_user.authenticity_token)
    .map_err(|err| {
        log::error!("Verify authenticity_token error: {}",
        err);
        Flash::error(
            Redirect::to("/users/new"),
            "Something went wrong when creating user",
        )
    })?;
```

If we continue logging the `verify()` method and continue tracing the source of the problem, we will eventually find out that the `from_request()` method of the token is not producing the correct result. We can fix the problem by changing the `from_request()` method in `src/fairings/csrf.rs` with the following lines:

```
async fn from_request(request: &'r Request<'_>) ->
Outcome<Self, Self::Error> {
    match request.get_csrf_token() {
        None => Outcome::Failure((Status::Forbidden, ())),
        Some(token) => Outcome::Success(Self(base64::
        encode_config(token, base64::URL_SAFE))),
    }
}
```

Obviously, logging the code and finding the problem is not efficient. We can also use a debugger such as `gdb` (GNU Debugger) or `lldb` to debug Rust programs. `gdb` can be used on the Linux operating system, and `lldb` (Debugger of LLVM Project) can be used on macOS and Linux operating systems. Please install one of those debuggers if you want to use a debugger for the Rust programming language.

Rust provides `rust-gdb` (a wrapper to `gdb`) and `rust-lldb` (a wrapper to `lldb`). Those programs should be installed with the Rust compiler. Let's see an example of how to use `rust-gdb` by following these steps:

1. First, build the application by using the `cargo build` command on the Terminal. Since we are not building the release version, debugging symbols should be in the resulting binary.

2. Check the location of the generated binary in the source directory target directory; for example, if the source code for the application is in `/workspace/rocketbook/Chapter12/03RocketTesting/`, we can find the generated binary in `/workspace/rocketbook/Chapter12/03RocketTesting/target/debug/our_application`.

3. Run `rust-gdb` on the Terminal just like you would run `gdb`. Here is an example:

```
rust-gdb -q target/debug/our_application
```

4. You'll see a gdb prompt as follows:

    ```
    Reading symbols from target/debug/our_application...
    Reading symbols from /workspace/rocketbook/
    Chapter12/03RocketTesting/target/debug/our_application/
    Contents/Resources/DWARF/our_application...
    (gdb)
    ```

5. Set the breakpoint of the application, as in this example:

    ```
    b /workspace/rocketbook/Chapter12/03RocketTesting/src/
    lib.rs:143
    ```

6. You'll see a prompt to set a breakpoint on the our_application library
    as follows:

    ```
    No source file named /workspace/rocketbook/
    Chapter12/03RocketTesting/src/lib.rs.
    Make breakpoint pending on future shared library load? (y
    or [n])
    ```

7. Reply with y and notice Breakpoint 1 set as follows:

    ```
    (y or [n]) y
    Breakpoint 1 (/workspace/rocketbook/
    Chapter12/03RocketTesting/src/lib.rs:143) pending.
    ```

8. Run the application by writing the r command and pressing the *Enter* key on the
    gdb prompt:

    ```
    (gdb) r
    ```

9. The application should run and because it hit the breakpoint, the execution stopped.
    We can use the gdb prompt again to inspect final_rocket, as in the following example:

    ```
    Starting program: /workspace/rocketbook/
    Chapter12/03RocketTesting/target/debug/our_application
      [Thread debugging using libthread_db enabled]
      Using host libthread_db library
    ```

```
"/usr/lib/libthread_db.so.1".
[New Thread 0x7ffff7c71640 (LWP 50269)]
...
[New Thread 0x7ffff746d640 (LWP 50273)]

Thread 1 "our_application" hit Breakpoint 1, our_
application::main::{generator#0} () at
src/lib.rs:143
143             final_rocket
```

10. Try printing some variables in the debugger prompt:

    (gdb) p config

    We can see the result printed as follows:

    ```
    $1 = our_application::Config {databases: our_
    application::Databases {main_connection: our_
    application::MainConnection {url: "postgres://
    username:password@localhost/rocket"}}, jwt_secret: "+/
    xbAZJs+e1BA4
    gbv2zPrtkkkOhrYmHUGnJIoaL9Qsk="}
    ```

11. To quit from gdb, just type quit on prompt and confirm quitting the debugger as follows:

    ```
    (gdb) quit
    A debugging session is active.

            Inferior 1 [process 50265] will be killed.

    Quit anyway? (y or n) y
    ```

There are many more functionalities in these debuggers, such as setting multiple breakpoints and stepping through the breakpoints. You can find more information about gdb at https://www.sourceware.org/gdb/, and about lldb at https://lldb.llvm.org/.

There are also debuggers for IDEs or code editors, for example, users of Visual Studio Code can use CodeLLDB (`https://marketplace.visualstudio.com/items?itemName=vadimcn.vscode-lldb`) to conveniently click on the line and mark the breakpoints from the editor and inspect the variables through a dedicated panel:

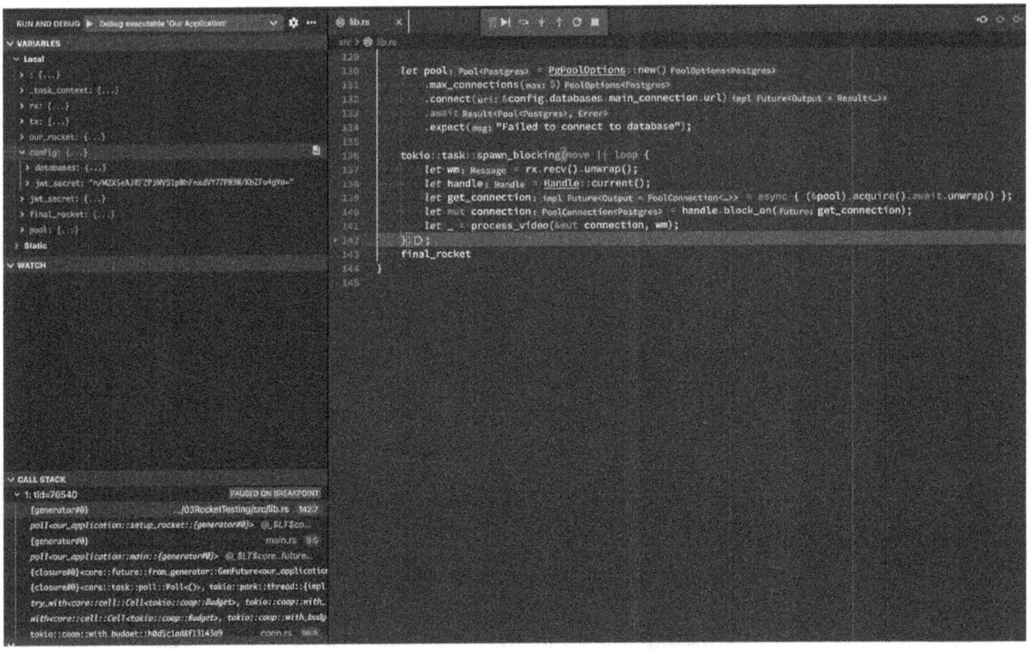

Figure 12.1 – CodeLLDB inspecting our application

In any case, using a debugger is an indispensable tool for programming. Learning to use debuggers properly can help in working with the Rust programming language.

Summary

In this chapter, we learned about testing the Rust program and Rocket application. We learned about using macros such as `assert!` to do the testing. We also learned the difference between unit testing and functional testing.

We created a functional test and learned about modules to do functional testing on a Rocket application. Finally, we learned a little bit about the technique to debug a Rust application to help fix it.

Testing and debugging are important parts of programming, as these techniques can improve the correctness of the application.

After all the development is done, in the next chapter, we are going to learn several ways to set the Rocket application available to serve real-world users.

13
Launching a Rocket Application

After development and testing, an important part of development is preparing the application to serve its intended users. In this chapter, we are going to learn some techniques to generate a production-ready binary. After we generate the binary, we are going to learn about configuring the application behind a general-purpose web server. And finally, we will learn how to generate Docker images for a Rocket application.

After learning the information in this chapter, you will be able to optimize binaries using Rust compiler flags and Cargo configurations. You will also learn techniques to prepare your applications to serve their intended users.

In this chapter, we are going to cover these main topics:

- Optimizing production binaries

- Setting up the Apache HTTP Server with a Rocket application

- Generating Docker images for a Rocket application

Technical requirements

In this chapter, we are going to serve HTTP requests using the Apache HTTP Server (`https://httpd.apache.org/`). If you have a Unix-based operating system, you can usually find the Apache HTTP Server in your operating system package manager. If you have a Windows operating system, there are recommended downloads at the following link: `https://httpd.apache.org/docs/2.4/platform/windows.html`.

You also need to generate a TLS (Transport Layer Security) certificate using OpenSSL. If you have a Unix-based operating system, you can usually find the OpenSSL binary using the distribution's package manager. If you have a Windows operating system, you can find the recommended binary at the following link: `https://wiki.openssl.org/index.php/Binaries`.

For generating Docker images, you can use Docker Desktop from the following link: `https://www.docker.com/products/docker-desktop/`.

You can find the source code of this chapter at `https://github.com/PacktPublishing/Rust-Web-Development-with-Rocket/tree/main/Chapter13`.

Optimizing production binaries

After we create the application, we want to prepare the application to accept a real connection. In software development, there's a production environment, also called a release environment or deployment environment. The production environment contains the configuration of the system and software to make it available to the intended customer. In *Chapter 2, Building Our First Rocket Web Application*, we learned that we can tell the Rust compiler to build release binary when compiling the Rust application. We can use `cargo build` or `cargo run` with the extra `--release` flag.

To refresh, Cargo will read the configuration in `Cargo.toml` in the `[profile.release]` section. There are some compilation optimizations we can do to improve the resulting image:

1. The first one is the number of `codegen-units` of compilation. Rust compilation can take significant time, and to solve this, the compiler may try to split it into parts and compile them in parallel. But, compiling binaries or libraries in parallel might omit some optimization. By default, the number of `codegen-units` is 3. We can sacrifice the compilation speed and make `codegen-units` into 1 to optimize the resulting binary further. For example, in `Cargo.toml`, we can have the following:

    ```
    [profile.release]
    codegen-units = 1
    ```

Link Time Optimization (LTO): The Rust codegen backend, LLVM, can perform various LTOs to produce output with optimized code. To enable LTO, we can set lto = yes or lto = "fat". The following is an example of lto in Cargo.toml:

```
[profile.release]
lto = "fat"
```

Set the optimization level. We can set optimization levels from 0, 1, 2, and 3, with the default value being 0 (no optimization) up to 3 (all optimization) as in the following:

```
[profile.release]
opt-level = 3
```

Besides optimization level 0 to 3, we can also also set "s" and "z", with "s" for binary size optimization and "z" for binary size optimization and turn of loop vectorization.

Disable panic unwinding. We can set panic to not show the stack trace. The result is a more optimized binary. Set the following in Cargo.toml to disable stack unwinding:

```
[profile.release]
panic = "abort"
```

2. The second optimization is compiling the correct architecture. CPU producers will always create a new CPU with better optimization or instruction sets that can improve the performance of an application. For example, the SSE (Streaming SIMD Extensions) instruction set was introduced by Intel with the release of Intel Pentium III.

    By default, the Rust compiler will produce binary with a reasonable amount of CPU support. But, this means newer instruction sets or optimizations are not used when compiling a library or a binary. We can tell the Rust compiler to produce a binary that supports newer optimizations or instruction sets of the target machine.

    To see the list of the target architecture supported by the Rust compiler, we can use this command:

    ```
    rustc --print target-cpus
    ```

For example, if we know the target machine is AMD Ryzen, which supports *znver3* architecture, we can compile the Rust program as follows:

```
RUSTFLAGS='-C target-cpu=znver -C codegen-units=1' cargo
build -release
```

The reason we use the `RUSTFLAGS` environment variable is that `target-cpu` is not a recognized option in `Cargo.toml`. Cargo will also use any other `rustc` option set in the `RUSTFLAGS` environment variable.

Now we know how to compile a Rust application for a production environment, let's learn about deploying a Rocket application behind a web server.

Setting up the Apache HTTP Server with a Rocket application

We know that Rocket has TLS support in its configuration, so we can set the TCP port to 443, the default HTTPS connection port. In some cases, it might be acceptable to run web applications directly, for example, when we want to serve content for microservices.

One of the reasons why we don't want to run the Rocket application directly is because of this warning in the Rocket guide:

> *Rocket's built-in TLS implements only TLS 1.2 and 1.3. It may not be suitable for production use.*

The TLS library used by the Rocket framework might not be suitable for production use for various reasons, such as security reasons or it is not yet audited.

There are other reasons why we do not want to serve content directly from Rocket aside from the TLS library problem. One example is when we want to serve multiple applications from a single computer. We might want to serve PHP applications from the same machine too.

One of the techniques people use when serving a Rust application is putting it behind a general-purpose web server that can do a reverse proxy:

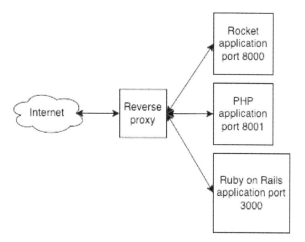

Figure 13.1 – General purpose web server performing a reverse proxy on a Rocket application

One of the most used reverse proxy applications is the Apache HTTP Server. The Apache HTTP Server also has other features besides the reverse proxy, including serving static files and compressing files to serve requests faster.

Let's try serving our application using the Apache HTTP Server and configuring the server to act as a reverse proxy by following these steps:

1. Download the Apache HTTP Server for your operating system or from `https://httpd.apache.org/`.

2. Try starting the application using the following command line:

    `sudo apachectl -k start`

3. The Apache HTTP Server's default port is `8080`. Check that Apache is running by using the cURL command:

    `curl http://127.0.0.1:8080/`

4. The Apache HTTP Server's functionalities can be extended by modules, and several modules are installed alongside it. We want to enable several modules to enable HTTP requests to our application using a reverse proxy. Find `httpd.conf`, the configuration file for your operating system. In some Linux distributions, the configuration might be in `/etc/httpd/httpd.conf`. In other distributions or operating systems, the file location might be in `/usr/local/etc/httpd/httpd.conf`.

Edit the `httpd.conf` file and remove the comment to enable the required modules:

```
LoadModule log_config_module libexec/apache2/mod_log_
config.so
LoadModule vhost_alias_module libexec/apache2/mod_vhost_
alias.so
LoadModule socache_shmcb_module libexec/apache2/mod_
socache_shmcb.so
LoadModule ssl_module libexec/apache2/mod_ssl.so
LoadModule xml2enc_module libexec/apache2/mod_xml2enc.so
LoadModule proxy_html_module libexec/apache2/mod_proxy_
html.so
LoadModule proxy_module libexec/apache2/mod_proxy.so
LoadModule proxy_connect_module libexec/apache2/mod_
proxy_connect.so
LoadModule proxy_http_module libexec/apache2/mod_proxy_
http.so
```

5. In the same `httpd.conf` file, find these lines and uncomment these lines as well:

```
Include /usr/local/etc/httpd/extra/httpd-vhosts.conf
Include /usr/local/etc/httpd/extra/httpd-ssl.conf
Include /usr/local/etc/httpd/extra/proxy-html.conf
```

6. We need a server name. In a real server, we can acquire a domain by buying the rights to it from a domain registrar and pointing the **domain A** record to the server IP address using domain registrar management tools. But, as we're developing in a development environment, we need a fake domain such as `ourapplication.example.net`. Edit `/etc/hosts` and some test domains as follows:

```
127.0.0.1 ourapplication.example.net
```

7. Install `openssl` for your operating system. After that, generate a certificate for `ourapplication.example.net` using the `openssl` command line, as in the following:

```
openssl req -x509 -out ourapplication.example.com.crt
-keyout ourapplication.example.com.key \
  -newkey rsa:2048 -nodes -sha256 \
  -subj '/CN=ourapplication.example.com' -extensions
```

```
EXT -config <( \
 printf "[dn]\nCN=ourapplication.example
.com\n[req]\ndistinguished_name = dn\n[EXT]\
nsubjectAltName=DNS:ourapplication.
example.com\nkeyUsage=digitalSignature\
nextendedKeyUsage=serverAuth")
```

The command line will generate two files, `ourapplication.example.com.crt` and `ourapplication.example.com.key`.

8. Generate a PEM file, a file format that contains certificate as follows:

```
openssl rsa -in ourapplication.example.com.key -text >
ourapplication.example.com.private.pem
openssl x509 -inform PEM -in ourapplication.example.com.
crt > ourapplication.example.com.public.pem
```

9. Edit `httpd-vhosts.conf`. The file might be in `/usr/local/etc/httpd/extra/`, depending on your operating system configuration. Add a new virtual host. We want the virtual host to point to our Rocket application at `http://127.0.0.1:8000`. Add the following lines:

```
<VirtualHost *:443>
      ServerName ourapplication.example.com
      SSLEngine On
      SSLCertificateFile /usr/local/etc/httpd/
      ourapplication.example.com.public.pem
      SSLCertificateKeyFile /usr/local/etc/httpd/
      ourapplication.example.com.private.pem
      SSLProxyEngine On
      ProxyRequests Off
      ProxyVia Off
      <Proxy *>
          Require all granted
      </Proxy>
      ProxyPass "/" "http://127.0.0.1:8000/"
      ProxyPassReverse "/" "http://127.0.0.1:8000/"
</VirtualHost>
```

10. Check whether the configuration is correct by running the following command:

```
sudo apachectl configtest
```

11. Restart and open `https://ourapplication.example.com` in your web browser. The web browser might complain because the root certificate is unknown. We can add our generated certificate so it's accepted in our browser. For example, in Firefox, we can go to **Preferences** | **Privacy & Security** | **View Certificates**. After that, choose **Servers Tab** and click **Add Exception**. Then, ensure that **Permanently store this exception** is checked. Finally, click on **Confirm Security Exception** to store the security exception. If everything goes well, we can use the example domain in the browser, as in the following figure:

Figure 13.2 – Using a domain and TLS certificate

Now that we have deployed the Rocket application behind a reverse proxy, we can use the same principle with a real server. Set up the Apache HTTP Server or NGINX as a reverse proxy and run the Rocket application behind the reverse proxy.

To run the Rocket application automatically when the operating system starts, we can set up some kind of service for the operating system. If we are running a Linux distribution with systemd as a service manager, for example, we can create a `systemd` service file and run the application automatically.

In the next section, we are going to learn a different way to deploy an application. We are going to use Docker to package and create a Docker container for our Rocket application.

Generating Docker images for a Rocket application

Containerization has been a popular choice to ship production applications for a while. One of the most popular applications for containerization is Docker. In this section, we are going to learn how to set up Docker to run our Rocket application. To use the `docker` command line, please install Docker Desktop from `https://www.docker.com/products/docker-desktop/`.

Follow these steps to create and run a Docker image of the Rocket application:

1. In the root folder of the application, create a Dockerfile.

2. There are some base images we can use to build and run the application. We are going to use Rust's official Docker image from `https://hub.docker.com/_/rust`. For the Linux distribution, we are going to use *Alpine base* because it's one of the smallest base images for Docker.

   In the Dockerfile, add the first line:

   ```
   FROM rust:alpine as prepare-stage
   ```

3. Set the working directory. Append this line to the Dockerfile:

   ```
   WORKDIR /app
   ```

4. We can use Cargo to install the dependencies, but there is another way to quickly compile the application. We can vendorize the dependencies and use the vendor dependencies to build the application. Run this command on the root folder of the application source code:

   ```
   cargo vendor
   ```

5. We want to override the source of the dependencies from the internet to the vendor folder. Create a `.cargo` folder in the root application folder, and create `config.toml` inside the `.cargo` folder.

   Append these lines to the `.cargo/config.toml` file:

   ```
   [source.crates-io]
   replace-with = "vendored-sources"

   [source.vendored-sources]
   directory = "vendor"
   ```

6. We want to add the required files to build the application as a Docker image. We don't need `Rocket.toml`, templates, or static files to build the application. Append these lines to the Dockerfile:

```
COPY src src
COPY Cargo.toml Cargo.toml
COPY .cargo .cargo
COPY vendor vendor
```

7. Add the instructions to build the image. We want to use another stage and install the dependencies to build the image. Add the following lines:

```
FROM prepare-stage as build-stage
RUN apk add --no-cache musl-dev
RUN cargo build --release
```

8. Try building the application by running the following command:

```
docker build .
```

9. After testing, add a new part to run the application in the Dockerfile. We want to open port `8000`. We also want to add a default time zone and configure the user to run the application. Append the following lines:

```
FROM rust:alpine
EXPOSE 8000
ENV TZ=Asia/Tokyo \
    USER=staff
RUN addgroup -S $USER \
    && adduser -S -g $USER $USER
```

10. We want the image to have the latest libraries. Append the following lines to the Dockerfile:

```
RUN apk update \
    && apk add --no-cache ca-certificates tzdata \
    && rm -rf /var/cache/apk/*
```

11. Set the working directory. Append the following line to the Dockerfile:

```
WORKDIR /app
```

12. Set `Rocket.toml` to run from `0.0.0.0`. We want to tell the application to use the host's running database. In Docker, we can reference the host machine using a special domain, `host.docker.internal`. Edit `Rocket.toml` as follows:

```
[default.databases.main_connection]
url = "postgres://username:passwordR@host.docker.
internal:5432/rocket"
[release]
address = "0.0.0.0"
```

13. Copy the resulting binary, `Rocket.toml`, assets, and templates to the final image. Append the following lines to the Dockerfile:

```
COPY --from=build-stage /app/target/release/our_
application our_application
COPY Rocket.toml Rocket.toml
COPY static static
COPY src/views src/views
```

14. Add the folder to store the log file:

```
RUN mkdir logs
```

15. Add changing permission to `$USER` as follows:

```
RUN chown -R $USER:$USER /app
```

16. Finally, run the entry point to the application to the Dockerfile:

```
USER $USER
CMD ["./our_application"]
```

17. Build the image and create a tag for it using this command:

```
docker build -t our_application .
```

18. After building the Docker image, it's time to run it. Use the following command line:

```
docker run --add-host host.docker.internal:host-gateway
-dp 8000:8000 our_application
```

After everything is done, we should see the Docker container is running and showing the `our_application` output:

Figure 13.3 – Docker Desktop showing the running container and our_application

Deploying a Rocket application using Docker is just like deploying other applications. We need to copy the source, build, and run the resulting image. There are some actions that we can perform to ensure proper deployment, such as vendoring the libraries and opening the correct ports to ensure requests can be made to the running container and applications running inside the container.

Summary

In this chapter, we have learned about production-ready compilation options. We can use them to ensure the resulting binary is as optimized as possible. We also learned about setting up a general-purpose HTTP server to work in conjunction with a Rocket application. And finally, we learned to create and run Docker images for the Rocket application.

After learning these techniques, we expanded them to set up the Rocket application to serve its intended users.

In the next chapter, we are going to learn about using Rust to create a frontend WebAssembly application in conjunction with a Rocket application.

14

Building a Full Stack Application

In this chapter, we are going to learn how to build a simple WebAssembly application and use Rocket to serve the WebAssembly application. We are going to make WebAssembly load the user information from one of the endpoints that we created earlier. After learning the information in this chapter, you will be able to write and build a WebAssembly application using Rust. You will learn how to serve WebAssembly using the Rocket web framework.

In this chapter, we are going to cover these main topics:

- Introducing WebAssembly
- Setting up a Cargo workspace
- Setting a WebAssembly build target
- Writing a WebAssembly application using Yew
- Serving a WebAssembly application using Rocket

Technical requirements

The technical requirements for this chapter are very simple: the Rust compiler, the Cargo command line, and a web browser.

You can find the code for this chapter at `https://github.com/PacktPublishing/Rust-Web-Development-with-Rocket/tree/main/Chapter14`.

Introducing WebAssembly

In the past, almost all applications in the web browser were made using the JavaScript language. There were also attempts to use different languages in the web browser, such as Java Applet, Adobe Flash, and Silverlight. But, all those different attempts were not web standards, and, therefore, the adoption of those attempts was not as universal as JavaScript.

However, there is a way to use other programming languages in the web browser: by using **WebAssembly**. WebAssembly is both a binary executable format and its corresponding text format for a stack-based virtual machine. Web browsers that support WebAssembly can execute the binary executable format. Any programming language that can be compiled into WebAssembly can be executed by web browsers.

In 2015, WebAssembly was announced and was first released in March 2017. All major web browser vendors finished the release of browsers with minimum support for WebAssembly in September 2017, and then World Wide Web Consortium recommended WebAssembly on December 5th, 2019.

Compiled languages such as C++ or Rust can be compiled into a `.wasm` file, and a virtual machine in the browser can then run the WebAssembly file. To run interpreted languages, first, the language runtime can be compiled into a `.wasm` file, and then, the runtime can run the runtime scripts.

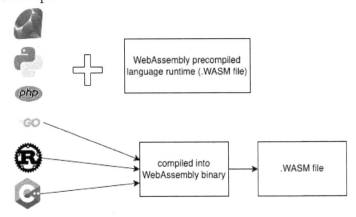

Figure 14.1 - Interpreted languages and compiled languages in WebAssembly

The Rust programming language supports WebAssembly, and as we have already learned about Rust and created a backend application using Rust and Rocket, we can take this opportunity to learn a little about frontend application development using Rust. The old web standards and web technologies, such as HTML, CSS, and JavaScript, are technologies that changed the course of human history. Learning about new web standards, such as WebAssembly, is a good opportunity to be a part of future development.

Let's implement a page in our application where we will render an empty template. The template will load WebAssembly binary from the server. WebAssembly will call the user API endpoint that we created earlier. It will then render users using a custom component.

For the implementation, we are going to use Yew (`https://yew.rs`), which is a frontend Rust framework.

Setting up a Cargo workspace

Since we are going to create a new application, it would be nice if we could make the code for the `our_application` Rocket application work alongside this new application. Cargo has a feature called **Cargo workspaces**. A Cargo workspace is a set of different Cargo packages in a single directory.

Let's set up a Cargo workspace to have multiple applications in a single directory using the following steps:

1. Create a directory, for example, `01Wasm`.

2. Move the `our_application` directory inside the `01Wasm` directory and create a new `Cargo.toml` file inside the `01Wasm` directory.

3. Edit the `Cargo.toml` file as follows:

    ```
    [workspace]

    members = [
      "our _ application",
    ]
    ```

4. Create a new Rust application inside `01Wasm` using this command:

    ```
    cargo new our _ application _ wasm
    ```

5. After that, add the new application as a member of the workspace in `01Wasm/Cargo.toml`, as follows:

    ```
    members = [
      "our_application",
      "our_application_wasm",
    ]
    ```

6. Try building both applications using this command:

    ```
    cargo build
    ```

7. To build or run one of the applications, append `--bin` with the binary package name, or `--lib` with the library package name. To run the application, consider the location of the directories required for running the Rocket application. For example, if there's no logs directory, the application might fail to run. Also, if there's no static directory, the application might not be able to find the assets file.

8. Try building one of the applications by running this command in the terminal:

    ```
    cargo build --bin our_application
    ```

Now that we have set up the Cargo workspace, we can learn how to build the application for a different target specifically for WebAssembly.

Setting a WebAssembly build target

The Rust compiler can be set to compile to a different architecture. The architectures are also called **targets**. Targets can be identified by using **target triple**, a string that consists of three identifiers to be sent to the compiler. Examples of targets are `x86_64-unknown-linux_gnu` and `x86_64-apple-darwin`.

Targets can be categorized into three tiers, tier 1, tier 2, and tier 3:

- **Tier 1** means that the target is guaranteed to work properly.

- **Tier 2** means that the target is guaranteed to build but, sometimes, the automated test to build the binary for the targets may not pass. The host tools and full standard library for this tier are also supported.

- **Tier 3** means that the Rust code base supports some features of the targets. The build for those targets may or may not exist, and the tooling might not be complete.

Remember that WebAssembly is a binary format for a virtual machine. The Rust compiler has targets for the virtual machine specifications, such as `asmjs-unknown-emscripten`, `wasm32-unknown-emscripten`, or `wasm32-unknown-unknown`. The community mostly supports the tooling around `wasm32-unknown-unknown`.

To see the available target list for the Rust compiler, run the following command in the terminal:

```
rustup target list
```

To add WebAssembly target support for the Rust compiler, run the following command in the terminal:

```
rustup target add wasm32-unknown-unknown
```

After adding the target, try building `our_application_wasm` by running this command:

```
cargo build --target wasm32-unknown-unknown --bin our_
application _ wasm
```

We will use `wasm32-unknown-unknown` to build the WebAssembly application in the next section.

Writing a WebAssembly application using Yew

In the application, we are going to use Yew (`https://yew.rs`). On the website, it says that Yew is a modern Rust framework for creating multithreaded frontend web apps.

Cargo can compile the WebAssembly binary, but the WebAssembly binary itself is not usable without other steps. We have to load the WebAssembly binary in the web browser's virtual machine engine. There are proposals such as using a `<script type="module"></script>` tag, but unfortunately, those proposals are not standard yet. We have to tell JavaScript to load the module using the WebAssembly Web API. To make the development easier, we can use `wasm-pack` from the Rust WebAssembly working group at `https://rustwasm.github.io/`. Yew uses an application named `trunk` (`https://trunkrs.dev`) that wraps `wasm-pack` and provides other conveniences. Install `trunk` by using this command:

```
cargo install --locked trunk
```

Now that the preparation for compiling WebAssembly has been completed, we can write the code for the WebAssembly application. Follow these steps to create the application:

1. Create an HTML file named index.html inside the our_application_wasm directory. We will use this HTML file to mimic the template on our_application, with small differences. We want to add an ID for an HTML tag to be the main tag for the WebAssembly application. Let's name this ID main_container. Append the following lines to our_application_wasm/index.html:

    ```
    <!DOCTYPE html>
    <html lang="en">

    <head>
    </head>

    <body>
      <header>
        <a href="/" class="button">Home</a>
      </header>
      <div class="container" id="main _ container"></div>
    </body>
    </html>
    ```

2. Add yew as a dependency to our_application_wasm. We also want to access the browser DOM, so we need another dependency. Gloo (https://gloo-rs.web.app/) provides bindings to the Web API, and we want to use gloo_utils as a dependency for our WebAssembly application to access the DOM. Add the following dependencies to our_application_wasm/Cargo.toml:

    ```
    gloo-utils = "0.1.3"
    yew = "0.19"
    getrandom = { version = "0.2", features = ["js"] }
    ```

3. Add the required use declarations in our_application_wasm/src/main.rs:

    ```
    use gloo _ utils::document;
    use yew::prelude::*;
    ```

4. Create a minimal component that creates an empty HTML inside our_
   application_wasm/src/main.rs:

```
#[function _ component(App)]
fn app() -> Html {
    html! {
        <>{"Hello WebAssembly!"}</>
    }
}
```

5. Use gloo_utils in the main() function in our_application_wasm/
   src/main.rs to select the div tag with the main_container ID. Append the
   following lines in the main() function:

```
let document = document();
let main _ container = document.query _ selector("#main _
container").unwrap().unwrap();
```

6. Initialize a Yew application by appending this line to the main() function:

```
yew::start _ app _ in _ element::<App>(main _ container);
```

7. We can use trunk to create a small web server and build everything needed to
   build the WebAssembly and related JavaScript to load the WebAssembly and serve
   the HTML. Run this command in the terminal inside the our_application_
   wasm directory:

   trunk serve

   There should be an output like the following in the terminal:

```
Apr 27 20:35:44.122   INFO fetching cargo artifacts
Apr 27 20:35:44.747   INFO processing WASM
Apr 27 20:35:44.782   INFO using system installed binary
app="wasm-bindgen" version="0.2.80"
Apr 27 20:35:44.782   INFO calling wasm-bindgen
Apr 27 20:35:45.065   INFO copying generated wasm-bindgen
artifacts
Apr 27 20:35:45.072   INFO applying new distribution
Apr 27 20:35:45.074   INFO ✅ success
Apr 27 20:35:45.074   INFO 🚴 serving static assets at -> /
```

```
Apr 27 20:35:45.075  INFO 🖧 server listening at 0.0.0.0:8080
Apr 27 20:53:10.796  INFO 📄 starting build
Apr 27 20:53:10.797  INFO spawning asset pipelines
Apr 27 20:53:11.430  INFO building our_application_wasm
```

8. Try opening a web browser to `http://127.0.0.1:8080`; you'll see that it loads and runs the Yew WebAssembly application:

Figure 14.2 – Hello WebAssembly!

9. We are going to get users' information using an API endpoint that returns the JSON that we created earlier in `our_application` from `http://127.0.0.1:8000/api/users`. To convert the JSON into Rust types, let's define similar types to those in `our_application`. The types should derive SerDes' `deserialize`. In `our_application_wasm/Cargo.toml`, add the dependencies for the WebAssembly code:

```
chrono = {version = "0.4", features = ["serde"]}
serde = {version = "1.0.130", features = ["derive"]}
uuid = {version = "0.8.2", features = ["v4", "serde"]}
```

10. Then, in `our_application_wasm/src/main.rs`, add the required `use` declarations:

```
use chrono::{offset::Utc, DateTime};
use serde::Deserialize;
use std::fmt::{self, Display, Formatter};
use uuid::Uuid;
```

11. Finally, add the types to deserialize the JSON:

```
#[derive(Deserialize, Clone, PartialEq)]
enum UserStatus {
    Inactive = 0,
    Active = 1,
}
```

```rust
impl fmt::Display for UserStatus {
    fn fmt(&self, f: &mut fmt::Formatter<'_>) ->
    fmt::Result {
        match *self {
            UserStatus::Inactive => write!(f,
            "Inactive"),
            UserStatus::Active => write!(f, "Active"),
        }
    }
}


#[derive(Copy, Clone, Deserialize, PartialEq)]
struct OurDateTime(DateTime<Utc>);

impl fmt::Display for OurDateTime {
    fn fmt(&self, f: &mut fmt::Formatter<'_>) ->
    fmt::Result {
        write!(f, "{}", self.0)
    }
}

#[derive(Deserialize, Clone, PartialEq)]
struct User {
    uuid: Uuid,
    username: String,
    email: String,
    description: Option<String>,
    status: UserStatus,
    created_at: OurDateTime,
    updated_at: OurDateTime,
}

#[derive(Clone, Copy, Deserialize, PartialEq)]
struct Pagination {
    next: OurDateTime,
```

```
        limit: usize,
    }


    #[derive(Deserialize, Default, Properties, PartialEq)]
    struct UsersWrapper {
        users: Vec<User>,
        #[serde(skip _ serializing _ if = "Option::is _ none")]
        #[serde(default)]
        pagination: Option<Pagination>,
    }
```

> **Note**
>
> One thing that we can do to improve redefining the types is to create a library that defines types that can be used by both applications.

12. If we look at the `User` struct, we can see that the description field is an `Option`. Create a convenience function to return an empty `String` if the value is `None`, and return the `String` content of `Some` if the value is `Some`. Add the following function to `our_application_wasm/src/main.rs`:

```
    struct DisplayOption<T>(pub Option<T>);


    impl<T: Display> Display for DisplayOption<T> {
        fn fmt(&self, f: &mut Formatter) -> fmt::Result {
            match self.0 {
                Some(ref v) => write!(f, "{}", v),
                None => write!(f, ""),
            }
        }
    }
```

13. It's now time to implement a component that will render `User`. We will name the component `UsersList`. Add the following function to `our_application_wasm/src/main.rs`:

```
    #[function _ component(UsersList)]
    fn users _ list(UsersWrapper { users, .. }: &UsersWrapper)
    -> Html {
        users.iter()
```

```
.enumerate().map(|user| html! {
<div class="container">
    <div><mark class="tag">{ format!("{}",
    user.0) }</mark></div>
    <div class="row">
        <div class="col-sm-3"><mark>{ "UUID:"
        }</mark></div>
        <div class="col-sm-9"> { format!("{}",
        user.1.uuid) }</div>
    </div>
    <div class="row">
        <div class="col-sm-3"><mark>{
        "Username:" }</mark></div>
        <div class="col-sm-9">{ format!("{}",
        user.1.username) }</div>
    </div>
    <div class="row">
        <div class="col-sm-3"><mark>{ "Email:"
        }</mark></div>
        <div class="col-sm-9"> { format!("{}",
        user.1.email) }</div>
    </div>
    <div class="row">
        <div class="col-sm-3"><mark>{
        "Description:" }</mark></div>
        <div class="col-sm-9"> { format!("{}",
        DisplayOption(user.1.description.
        as _ ref())) }</div>
    </div>
    <div class="row">
        <div class="col-sm-3"><mark>{
        "Status:" }</mark></div>
        <div class="col-sm-9"> { format!("{}",
        user.1.status) }</div>
    </div>
    <div class="row">
```

```
                <div class="col-sm-3"><mark>{ "Created
                At:" }</mark></div>
                <div class="col-sm-9"> { format!("{}",
                user.1.created _ at) }</div>
            </div>
            <div class="row">
                <div class="col-sm-3"><mark>{ "Updated
                At:" }</mark></div>
                <div class="col-sm-9"> { format!("{}",
                user.1.updated _ at) }</div>
            </div>
            <a href={format!("/users/{}",
            user.1.uuid)} class="button">{ "See user"
            }</a>
        </div>
    }).collect()
}
```

Notice that the `html!` macro content looks like the content of our_
application/src/views/users/_user.html.tera.

14. We want to load the `User` data from the API endpoint. We can do this by using
two libraries, `reqwasm` (which provides HTTP request functionality), and `wasm-
bindgen-futures` (which converts Rust `futures` into JavaScript `promise`
and vice versa). Add the following dependencies to our_application_wasm/
Cargo.toml:

```
reqwasm = "0.2"
wasm-bindgen-futures = "0.4"
```

15. In our_application_wasm/src/main.rs, add a `const` for our API
endpoint. Add the following line:

```
const USERS _ URL: &str = "http://127.0.0.1:8000/api/users";
```

16. Implement the routine to fetch the `User` data. Add the required `use` declaration:

```
use reqwasm::http::Request;
```

Then, append the following lines inside the app () function in our_
application_wasm/src/main.rs:

```
fn app() -> Html {
    let users_wrapper = use_state(|| UsersWrapper::
    default());
    {
        let users_wrapper = users_wrapper.clone();
        use_effect_with_deps(
            move |_| {
                let users_wrapper =
                users_wrapper.clone();
                wasm_bindgen_futures::spawn_
                local(async move {
                    let fetched_users_wrapper:
                    UsersWrapper = Request::get(
                    USERS_URL)
                        .send()
                        .await
                        .unwrap()
                        .json()
                        .await
                        .unwrap();
                    users_wrapper.set(fetched_
                    users_wrapper);
                });
                || ()
            },
            (),
        );
    }
}
```

17. Below the {} block under the `users_wrapper` fetching, set the value for `next` and `limit`. Append the following lines:

```
let users _ wrapper = use _ state(|| UsersWrapper::default());
{
    ...
}
let (next, limit): (Option<OurDateTime>, Option<usize>) =
if users _ wrapper.pagination.is _ some()
{
    let pagination = users _ wrapper.
    pagination.as _ ref().unwrap();
    (Some(pagination.next), Some(pagination.limit))
} else {
    (None, None)
};
```

18. Change the HTML from `Hello WebAssembly!` to show the proper `User` information. We want to use the `UsersList` component that we created earlier. Change the `html!` macro content into the following:

```
html! {
    <>
        <UsersList users = {users _ wrapper.
        users.clone()}/>
        if next.is _ some() {
            <a href={ format!("/users?
            pagination.next={}&pagination.limit={}",
            DisplayOption(next), DisplayOption(limit))
            } class="button">
                { "Next" }
            </a>
        }
    </>
}
```

19. Build the `our_application_wasm` WebAssembly and JavaScript by running this command in the terminal:

```
trunk build
```

The command should generate three files in the `dist` directory: `index.html`, a WebAssembly file with random name, and a JavaScript file with random name. The example of random WebAssembly and JavaScript file are `index-9eb0724334955a2a_bg.wasm` and `index-9eb0724334955a2a.js` in the `dist` directory.

At this point, we have successfully written and built a WebAssembly application. In the next section, we are going to learn how to serve a WebAssembly application using Rocket.

Serving a WebAssembly application using Rocket

In this section, we are going to serve the WebAssembly web application using the following steps:

1. To run the WebAssembly file in `our_application`, we have to modify `our_application` a little bit. First, copy the WebAssembly and the JavaScript from `our_application_wasm/dist` to the `our_application/static` directory.

2. Edit the template to be able to selectively use WebAssembly in `our_application/src/views/template.html.tera` as follows:

```
<head>
    ...
    {% block wasm %}{% endblock wasm %}
    <meta...>
</head>

<body>
    ...
    {% block wasmscript %}{% endblock wasmscript %}
</body>
```

3. Add a new template file named `our_application/src/views/users/`
    `wasm.html.tera`. Edit the file in order to make sure the HTML file loads the
    necessary WebAssembly and JavaScript file, and run the WebAssembly on the right
    DOM. Add the following lines:

    ```
    {% extends "template" %}

    {% block wasm %}
    <link rel="preload" href="/assets/index-9eb0724334955a2a_
    bg.wasm" as="fetch" type="application/wasm" crossorigin="">
    <link rel="modulepreload" href="/assets/index-
    9eb0724334955a2a.js">
    {% endblock wasm %}

    {% block body %}
    <div id="main_container"></div>
    {% endblock body %}

    {% block wasmscript %}
    <script type="module">import init from '/assets/index-
    9eb0724334955a2a.js';init('/assets/index-9eb0724334955a2a_
    bg.wasm');</script>
    {% endblock wasmscript %}
    ```

4. Add a new route handling function to load just the generated HTML. Add the
    following function in `our_application/src/routes/user.rs`:

    ```
    #[get("/users/wasm", format = "text/html")]
    pub async fn wasm() -> HtmlResponse {
        let context = context! {};
        Ok(Template::render("users/wasm", context))
    }
    ```

5. Finally, don't forget to load the route. Add the new route in `our_application/`
    `src/lib.rs`:

    ```
    user::delete_user_entry_point,
    user::wasm,
    post::get_post,
    ```

6. Run the `our_application` web server by running `cargo run` in the
    `our_application` directory, and then open `http://127.0.0.1:8000/`
    `users/wasm` in the web browser. If we inspect the web browser developer tools,
    we can see that the web browser runs the JavaScript and WebAssembly, as shown in
    the following screenshot:

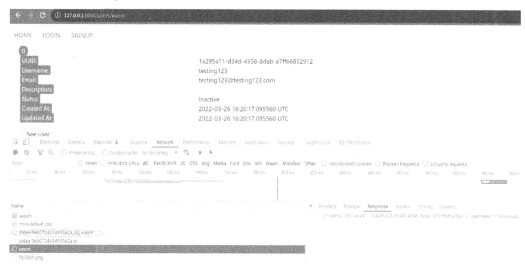

Figure 14.3 – Web browser loading and running our_application_wasm

WebAssembly should run properly by modifying the tag with the `main_container` tag,
then loading the JSON from `http://127.0.0.1:8000/api/users` and rendering
the HTML in the web browser properly.

Summary

Web technology has evolved to allow web browsers to run a universal binary format for a
virtual machine. Web browsers can now run a binary generated by the Rust compiler.

In this chapter, we have looked at an overview of WebAssembly, and how to prepare
the Rust compiler to compile to WebAssembly. We also learned how to set up a Cargo
workspace to have more than one application in a single directory.

We then learned how to write a simple frontend application that loads the `User` data
from the `our_application` API endpoint that we created earlier using Yew and
other Rust libraries.

Finally, we finished with how to serve the generated WebAssembly and JavaScript in the
`our_application` web server.

The next chapter is the final chapter, and we're going to see how we can expand the Rocket
application and find alternatives to it.

15
Improving the Rocket Application

Now that we have finished the simple application, in this final chapter, we will explore the improvements we can make to the Rocket application.

In this chapter, we will learn about adding various technologies such as logging, tracing, and monitoring to bring the Rocket application up to the standard of modern web development. We will explore techniques to scale the Rocket application.

We will also explore other web frameworks for the Rust language. One web framework might not be the best tool for everything, so by knowing about other web frameworks, we can broaden our knowledge of the Rust web ecosystem.

In this chapter, we are going to cover these main topics:

- Extending the Rocket application
- Scaling the Rocket application
- Exploring alternative Rust web frameworks

Technical requirements

The technical requirements for this chapter are very simple: the Rust compiler, the Cargo command line, and a web browser.

You can find the code for this chapter at `https://github.com/PacktPublishing/Rust-Web-Development-with-Rocket/tree/main/Chapter15`.

Extending the Rocket application

We have successfully created a simple Rocket application from scratch, starting with a basic Rocket concept such as routing. There are a lot of things that can be done to improve the application. In this section, we are going to discuss some of the libraries we can use to add functionality and improvements to the system.

Adding logging

A good web application in a modern setup usually requires logging and monitoring systems to obtain information about the system itself. Previously, we learned how to add logging to the Rocket application. The logging system writes to `stdout` and to a file. We can improve the logging system by using a distributed logging system in which the application sends the log to another server to create an ongoing record of application events.

We can create a Rocket fairing that sends log events to a third-party logging server such as Logstash, Fluentd, or Datadog. The logs can then be extracted, transformed, aggregated, filtered, and searched for further analysis.

An example of a crate that can be used to send a log to Fluentd is at `https://github.com/tkrs/poston`. Using a `poston` crate, we can create a worker pool to send data periodically to a Fluentd server.

Expanding logging to tracing

After setting logging for the Rocket application, we can improve the logging functionality further with the tracing concept. Where logging is usually concerned with recording an individual event, tracing is concerned with the workflow of an application. There are several terminologies that are commonly used, including **log**, **event**, **span**, and **trace**.

A **log** is a single piece of information used by programmers to capture data, while an **event** is the structured form of a log. For example, let's say we have a log using the **Common Log Format** (https://en.wikipedia.org/wiki/Common_Log_Format) as shown here:

```
127.0.0.1 user-identifier frank [10/Oct/2000:13:55:36 -0700]
"GET /apache_pb.gif HTTP/1.0" 200 2326
```

We can convert the log into an event as follows:

```
{
    "request.host": "127.0.0.1",
    "request.ident": "user-identifier",
    "request.authuser": "frank",
    "request.date": "2000-10-10 13:55:36-07",
    "request.request": "GET /apache_pb.gif HTTP/1.0",
    "request.status": 200,
    "request.bytes": 2326,
}
```

A **span** is a type of log, but instead of information from a single point in time, a span covers a duration. And, finally, a **trace** is a collection of spans that can be used to create a workflow of application parts.

Suppose we have a Rocket application with a fairing named `Trace`, we can implement tracing by using the `Trace` fairing and following these steps:

1. Create a struct that implements a Rocket request guard, for example, `RequestID`.
2. When a request arrives, the `Trace` fairing assigns `request_id` (an instance of `RequestID`) to the `Request` instance.
3. The `Trace` fairing then creates a log with the `request_id` and `start_time` information.
4. A route handling function then retrieves `request_id` as a parameter because the struct implements the Rocket request guard.
5. Inside the route handling function, the first thing we want the application to do is to create a log with `request_id` and the `function_start_time` information.
6. We can add various logs inside the function to record the timing; for example, before we send a query to the database, we create a log with `request_id` and time information. Later, we can create a log again when we receive the response from the database.

7. We can then add a log again before the function returns with the `request_id` and time to mark the end of the function.

8. Finally, in the `Trace` fairing, we create a log again with `request_id` and `end_time`.

By transforming and analyzing the logs, we can construct the logs with the same `request_id` into spans. Finally, we can construct the trees of the spans into a trace that records the timing of each event of a request in the Rocket request-response life cycle. By using tracing information, we can determine which parts of the application can be improved further.

There are a couple of crates that we can use to do the tracing, for example, `https://docs.rs/tracing/latest/tracing/` and `https://docs.rs/tracing-log/latest/tracing_log/`, which bridge the Rust logging functionality to the tracing functionality.

Setting monitoring

Where logging and tracing are used to obtain information for the Rocket application, then monitoring is the process to obtain information for the system to evaluate the system's capabilities itself. For example, we collect our server CPU usage for the Rocket application.

For monitoring, we can use tools such as Prometheus with Grafana as the visualizer, Datadog, or other third-party applications. We usually install an agent, an application that collects and sends various system information to a distributed monitoring server.

Even though there's no direct connection to the Rocket application, usually, a monitoring system also collects information about the application itself. For example, in a containerized environment, there are liveness and readiness concepts that make sure a container is ready to receive its intended function.

We can set a route in the Rocket application that returns a 200 HTTP status code, or a route that pings the database and returns a 200 HTTP status code. We can then tell the monitoring system to periodically check the response from the Rocket application. If there's a response, it means the application still works correctly, but if there's no response, it means there's something wrong with the Rocket application.

Setting up a mailing and alerting system

Sometimes, we need mailing functionality in a web application. For example, when a user registers on a website, the system then sends an email for verification. There are a couple of libraries to send an email for Rust. An example is the Lettre crate (`https://crates. io/crates/lettre`). Let's take a look at the sample code for sending an email.

In `Cargo.toml`, add the following dependencies:

```
[dependencies]
lettre = "0.9"
lettre_email = "0.9"
```

In the application, for example in `src/lib.rs`, we can add the following function to send an email:

```
use lettre::{SmtpClient, Transport};
use lettre_email::EmailBuilder;

fn send_email(email: &str, name: &str) -> Result<String,
String> {
    let email = EmailBuilder::new()
        .to((email, name))
        .from("admin@our_application.com")
        .subject("Hi, welcome to our_application")
        .text("Hello, thank you for joining our_
        application.")
        .build()
        .unwrap();

    let mut mailer = SmtpClient::new_unencrypted_
    localhost().unwrap().transport();
    mailer
        .send(email.into())
        .map(|_| String::from("Successfuly sent email"))
        .map_err(|_| String::from("Couldn't send email"))
}
```

One more thing we can add to the application is an alerting system for when something goes wrong. We can use a third-party notification system or use the mailing system to send a notification if something goes wrong.

Now that we have looked at several ways to improve the Rocket application, let's scale our application.

Scaling the Rocket application

After developing the Rocket application and deploying it to a production environment, the application might need to be scaled up due to increasing usage. There are a couple of ways to scale the web application, and they can be categorized into two categories: vertical scaling and horizontal scaling.

Vertical scaling means increasing the resources for a single node. For example, we replace the CPU of the computer that runs the Rocket application with a CPU with higher speed. Another example of vertical scaling is increasing the amount of RAM in the computer that runs the Rocket application.

Horizontal scaling is scaling the application by adding more nodes or more computers to handle the workload. An example of horizontal scaling is running two servers and setting up a Rocket web server on each server.

Suppose we have the following system:

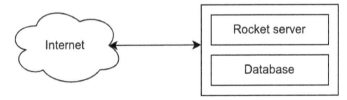

Figure 15.1 – Simple Rocket application

We can first move the database to another server as follows:

Figure 15.2 – Separating the database

Then, we can add a load balancer as in the following:

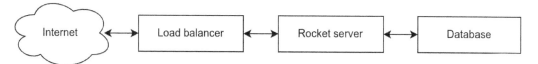

Figure 15.3 – Adding a load balancer

The load balancer can be a hardware load balancer, an IaaS (Infrastructure as a service) load balancer such as AWS Load Balancer, a Kubernetes load balancer, or a software load balancer such as HAProxy or NGINX.

After we add the load balancer, we can then add other machines, each with their own Rocket server instance, as follows:

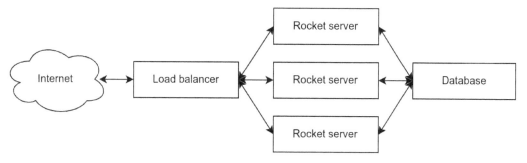

Figure 15.4 – Scaling the Rocket application horizontally

There are a couple of things we need to take care of if we want to load balance the Rocket server, for example, make sure "secret_key" in Rocket.toml is the same for all the Rocket server instances. Another thing we can do is make sure our session libraries and cookies are not storing the content on the memory of each instance, but on shared storage, such as a database.

Yet another idea to improve Rocket application scaling is hosting static files or assets on their own server. The static files server can be a general-purpose web server such as an Apache HTTP Server or NGINX or a service such as AWS S3 or Azure Storage. One thing we need to take care of is that when generating a Rocket response, we need to set the static assets into the correct server. For example, instead of setting the HTML CSS to "./mini-default.css", we have to set "https://static.example.com/mini-default.css".

A diagram of the static server along with the load balancer can be seen in the following figure:

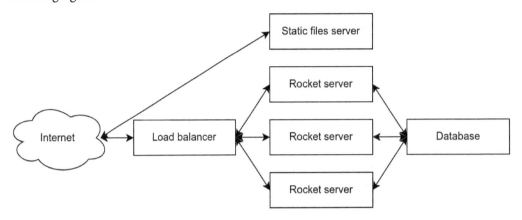

Figure 15.5 – Adding a static files server

We can also add a **content delivery network (CDN)** to distribute the load on the system, as follows:

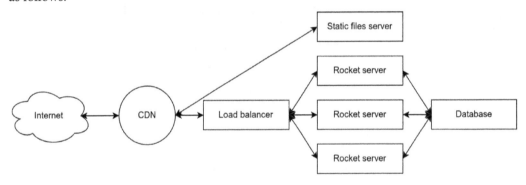

Figure 15.6 – Adding a CDN

The CDN can be from the IaaS, such as AWS CloudFront or GCP Cloud CDN, or a third-party CDN provider such as Fastly, Akamai, or Cloudflare. These CDNs provide servers in various geographical locations and can provide caching and a faster network connection to make our application faster.

After basic scaling, the system can be scaled further, such as by adding database replications or clusters, or adding caching systems such as Redis or Redis cache clusters. An example of such a system is shown here:

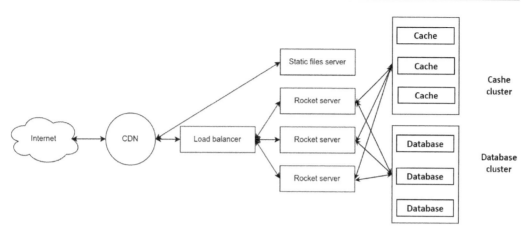

Figure 15.7 – Adding a database cluster and cache cluster

An important part of scaling the system is identifying which part's specification can be improved or which part can be isolated into its own server, for example, increasing the CPU of the computer that runs the Rocket server or moving the database into its own server and then later scaling the database itself from a single server into a database cluster.

Now that we have learned the basic techniques of scaling a Rocket application, let's discuss some other software that's similar to the Rocket web framework in the next section.

Exploring alternative Rust web frameworks

Rocket is a good web framework for the Rust programming language, but sometimes, we require other tools to build a web application. In this section, we are going to explore some alternatives to the Rocket web framework. The alternative frameworks are Actix Web, Tide, and Warp. Let's check the web frameworks one by one.

Actix Web

One good alternative to Rocket is Actix Web (`https://actix.rs/`). Just like Rocket, Actix Web is a web framework. Originally, it was created on top of the Actix crate, an actor framework. These days, the functionality from Actix is not used anymore as Rust's futures and `async/await` ecosystem is maturing.

Just like Rocket, Actix Web includes concepts such as routing, request extractor, form handler, response handler, and a middleware system. Actix Web also provides conveniences such as a static file handler, a database connection, templating, and many more.

Let's take a look at a code sample for Actix Web to see the similarities with Rocket.

In `Cargo.toml`, add the following:

```
[dependencies]
actix-web = "4.0.1"
```

And, in `src/main.rs`, add the following:

```
use actix_web::{get, web, App, HttpServer, Responder};

#[get("/users/{name}")]
async fn user(name: web::Path<String>) -> impl Responder {
    format!("Hello {name}!")
}

#[actix_web::main]
async fn main() -> std::io::Result<()> {
    HttpServer::new(|| {
        App::new()
            .route("/hello_world", web::get().to(|| async {
            "Hello World!" }))
            .service(user)
    })
    .bind(("127.0.0.1", 8080))?
    .run()
    .await
}
```

Try running the application and opening `http://127.0.0.1:8080/hello_world` or `http://127.0.0.1:8080/users/world` to see the result.

Tide

Another Rust web framework alternative is Tide (`https://github.com/http-rs/tide`). Unlike Rocket or Actix Web, this framework provides only basic functions, such as request type, result type, sessions, and middleware.

Let's take a look at a code sample for Tide to see the similarities with Rocket.

In `Cargo.toml`, add the following:

```
[dependencies]
tide = "0.16.0"
async-std = { version = "1.8.0", features = ["attributes"] }
```

And, in `src/main.rs`, add the following:

```
use tide::Request;

async fn hello_world(_: Request<()>) -> tide::Result {
    Ok(String::from("Hello World!").into())
}

#[async_std::main]
async fn main() -> tide::Result<()> {
    let mut app = tide::new();
    app.at("/hello_world").get(hello_world);
    app.listen("127.0.0.1:8080").await?;
    Ok(())
}
```

Try running the application by running `cargo run` on the command line and opening `http://127.0.0.1:8080/hello_world` in the browser.

Warp

Another Rust web framework alternative is Warp (`https://github.com/seanmonstar/warp`). This framework provides various functionalities on top of its filter function. By using the filter, it can perform path routing, extract parameters and headers, deserialize query strings, and parse various request bodies such as forms, multipart form data, and JSON. Warp also supports serving static files, directories, WebSocket, logging, middleware, and a basic compression system.

Let's take a look at an example application using Warp. In the `Cargo.toml` file, add the following:

```
[dependencies]
tokio = {version = "1", features = ["full"]}
warp = "0.3"
```

And, in the `src/main.rs` file, add the following:

```
use warp::Filter;

#[tokio::main]
async fn main() {
    let hello = warp::path!("hello_world")
        .and(warp::path::end())
        .map(|| format!("Hello world!"));

    warp::serve(hello).run(([127, 0, 0, 1], 8080)).await;
}
```

Again, like the Tide and Warp examples, try opening `http://127.0.0.1:8080/hello_world` in the browser.

Summary

In this final chapter, we have learned how to improve and scale a Rocket application. We can use various tools to improve Rocket applications, such as adding logging, tracing, monitoring, and mailers. We also learned a little bit about principles for scaling Rocket applications.

Finally, we learned about alternative Rust web frameworks such as Actix Web, Tide, 13 and Warp.

We started this book by learning how to create and build Rust application, and tools for working with Rust such as Cargo. We then learned the basics of Rocket applications such as the life cycle of requests and how to configure a Rocket application.

We then continued by learning about more concepts such as Rocket routes, and route parts such as HTTP method, path, format, and data. To handle a route, we have to create a function that receives the request object and returns the response object.

Continuing the basics of Rocket, we learned more about Rocket components such as state, connecting a database with Rocket, and fairings.

After that, we learned how to organize Rust modules to create more complex applications. We then designed an application and implemented the routes to manage entities such as user and post. To manage entities, we learned how to write queries to the database to add, get, modify, or delete items.

We then discussed more advanced topics such as Rust error handling and its implementation in a Rocket application. Continuing the more advanced topics, we also learned about Rocket functionalities such as serving static assets and using templates to generate a response. We also discussed how to use forms and how to protect forms from malicious attackers using CSRF.

After learning how to handle form data, we learned about the Rust generic and how to apply the Rust generic in a Rocket application to render `Post` with the same trait. To handle the variants of `Post`, we learned more about advanced Rust programming, including lifetime and memory safety. We also learned more about `async` programming and multithreading when we implemented processing of the variants of `Post`.

In order to use Rocket as a modern web framework, we also learned how to allow a Rocket application to handle APIs and JSON, protect the application using authentication and authorization, and learned how to use JWT to secure an API.

To make sure our Rocket application worked as intended, we then learned how to test Rust and Rocket applications. After making sure the application worked as intended, we learned how to deploy the application in different ways, such as putting the Rocket application behind a general-purpose web server and using Docker to build and serve the Rocket application.

To complement the backend application, we learned how to use Rust to create a WebAssembly application in the frontend. And finally, we learned more about how to scale Rocket applications, as well as how to find alternatives to the Rocket web framework.

Now that we have learned all the foundations for building Rust and Rocket applications, we can implement Rust and Rocket web framework skills in production-grade web applications. To expand upon your knowledge from this book, you can learn more from the Rust or Rocket websites and forums. Don't hesitate to experiment and make great applications using the Rust language and Rocket web framework.

Index

`Packt.com`

Subscribe to our online digital library for full access to over 7,000 books and videos, as well as industry leading tools to help you plan your personal development and advance your career. For more information, please visit our website.

Why subscribe?

- Spend less time learning and more time coding with practical eBooks and Videos from over 4,000 industry professionals
- Improve your learning with Skill Plans built especially for you
- Get a free eBook or video every month
- Fully searchable for easy access to vital information
- Copy and paste, print, and bookmark content

Did you know that Packt offers eBook versions of every book published, with PDF and ePub files available? You can upgrade to the eBook version at `packt.com` and as a print book customer, you are entitled to a discount on the eBook copy. Get in touch with us at `customercare@packtpub.com` for more details.

At `www.packt.com`, you can also read a collection of free technical articles, sign up for a range of free newsletters, and receive exclusive discounts and offers on Packt books and eBooks.

Other Books You May Enjoy

If you enjoyed this book, you may be interested in these other books by Packt:

Rust Web Programming

Maxwell Flitton

ISBN: 978-1-80056-081-9

- Structure scalable web apps in Rust in Rocket, Actix Web, and Warp
- Apply data persistence for your web apps using PostgreSQL
- Build login, JWT, and config modules for your web apps
- Serve HTML, CSS, and JavaScript from the Actix Web server
- Build unit tests and functional API tests in Postman and Newman

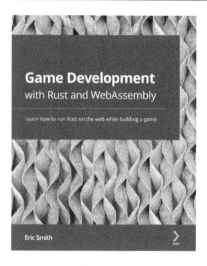

Game Development with Rust and WebAssembly

Eric Smith

ISBN: 978-1-80107-097-3

- Build and deploy a Rust application to the web using WebAssembly
- Use wasm-bindgen and the Canvas API to draw real-time graphics
- Write a game loop and take keyboard input for dynamic action
- Explore collision detection and create a dynamic character that can jump on and off platforms and fall down holes
- Manage animations using state machines

Packt is searching for authors like you

If you're interested in becoming an author for Packt, please visit authors. packtpub.com and apply today. We have worked with thousands of developers and tech professionals, just like you, to help them share their insight with the global tech community. You can make a general application, apply for a specific hot topic that we are recruiting an author for, or submit your own idea.

Share Your Thoughts

Now you've finished *Rust Web Development with Rocket*, we'd love to hear your thoughts! Scan the QR code below to go straight to the Amazon review page for this book and share your feedback or leave a review on the site that you purchased it from.

https://packt.link/r/180056130X

Your review is important to us and the tech community and will help us make sure we're delivering excellent quality content.